Jaya

Priangan woman with umbrella. (From J. Z. Van Dyck, *Garoet en Omstreken: Zwerftochten door der Preanger*, Batavia: G. Kolff and Co., 1922)

Java

A Travellers' Anthology

Compiled and Introduced by

JAMES R. RUSH

KUALA LUMPUR

OXFORD UNIVERSITY PRESS

OXFORD SINGAPORE NEW YORK

1996

Oxford University Press

Oxford New York
Athens Auckland Bangkok Bombay
Calcutta Cape Town Dar es Salaam Delhi
Florence Hong Kong Istanbul Karachi
Madras Madrid Melbourne Mexico City
Nairobi Paris Shah Alam Singapore
Taipei Tokyo Toronto

and associated companies in
Berlin Ibadan

Oxford is a trade mark of Oxford University Press

Published in the United States
by Oxford University Press, New York

First published 1996

British Library Cataloguing in Publication Data
Data available

Library of Congress Cataloging in Publication Data
Java: a traveller's anthology/compiled and introduced by James R. Rush.
p. cm
Includes bibliographical references.
ISBN 967 65 3082 4 (pbk.):
1. Java (Indonesia)—Description and travel. 2. Travelers—Indonesia—Java. I. Rush, James R. (James Robert), 1944–
DS646.19.J38 1996
915.98'204—dc 20
95–20614
CIP

Typeset by Indah Photosetting Centre Sdn. Bhd., Malaysia
Printed by KHL Printing Co. Pte. Ltd., Singapore
Published by the South-East Asian Publishing Unit,
a division of Penerbit Fajar Bakti Sdn. Bhd.,
under licence from Oxford University Press,
4 Jalan U1/15, Seksyen U1, 40000 Shah Alam,
Selangor Darul Ehsan, Malaysia

For Meg and Billy,
travellers too.

Preface

Coutus is Koedoes is Kudus

WESTERNERS have been writing about Java for several centuries, and this has provided them ample opportunity to mimic the sounds of local words and place-names in their own alphabet and spelling systems. The early English and Dutch visitors to Java were quite irregular spellers. Later arrivals, who were better spellers, applied the spelling systems of their own day; in the Dutch case, these systems were occasionally improved upon. Moreover, the Indonesian language itself has long since been rendered in the Western alphabet and also subjected to spelling reforms. As a result, the spelling of Indonesian place-names has changed dramatically over the centuries.

Thus, what early English travellers rendered as Sorobay had become Soerabaja on Dutch maps of the nineteenth century, by which time the English were writing Sourabaya. The Republic of Indonesia inherited the spelling Soerabaja but then, in stages, adopted Surabaja and then Surabaya, the current spelling. Similarly, today's Borobudur has been rendered Boro Bodo, Boro Boedor, and Boroboedoer at one time or another. Djocjakarta is Jogjakarta is Yogyakarta—and so on. Moreover, the Dutch renamed Bogor, Buitenzorg, and Jakatra, Batavia; after independence, Bogor became Bogor again, and Jakatra became Djakarta and, finally, Jakarta. Particularly vexing are the Preangan names with the English 'ch' sound, rendered with 'tj' in Dutch and the letter 'c' in contemporary Indonesian. Thus, Tjitjalengka of the 1930s has become Cicalengka of the 1990s; Tjitjoeroek is now Cicuruk; Tjiandjoer, Cianjur.

I have not altered the authentic spellings in the passages that follow, but I have, often, inserted the contemporary spelling of certain place-names in brackets so that readers can locate them on a modern map. The same device is used to gloss non-English words that our travellers have dropped into their accounts; the most frequently used terms, however, are listed in the Glossary below.

In compiling an anthology like this one, the choice of selections is invariably arbitrary. I have taken the opportunity to include some of my personal favourites gleaned from years of reading about Java—those by Alfred Russel Wallace, for example, and Augusta de Wit. Others I came across more recently while browsing in the library, like Eliot Elisofon's 1967 diary of a photo-shoot in Udjong Kulon wildlife sanctuary. Had I not read *They Went* by William Zinnzer, I might never have discovered David Fairchild's lovely memoirs of botanical wanderings in South-East Asia, or known that S. J. Perelman had visited Java in 1949. Similarly, my Arizona State University colleague, Richard Burg, introduced me to John Krich's 1980s romp through Asia, *Music in Every Room.* I wish to thank Anthony Reid for suggesting I take on this project, and Hermawan Sulistyo for assisting in background research. My wife, Sunny, put in many hours as volunteer editor and was generally invaluable. Finally, I thank the staff of the South-East Asian Publishing Unit of Oxford University Press who have deftly and patiently guided the arrival of this anthology.

Arizona State University JAMES R. RUSH
Tempe
April 1995

Acknowledgements

FOR permission to reproduce passages in this book, the Compiler and the Publishers gratefully acknowledge the following: Bill Berger Associates (Passage 29); Arthur Goodfriend (28); Harold Ober Associates (26); the Hakluyt Society (2, 3); McGraw-Hill (34); Alfred A. Knopf, and Aitken, Stone, and Wylie (33); Simon and Schuster (15, 21, 24, 25, 31); Walker and Company (32); and William Morrow and Company (30).

The Compiler and Publishers also wish to thank Richard Critchfield and the Universities Field Staff International, Maslyn Williams, Random House, Yale University Press, Charlotte Stryker, and the estate of H. M. Tomlinson.

Full citations and copyright acknowledgements are given following each passage.

The Compiler has made every effort to trace the copyright for passages in this anthology. In one or two cases these efforts have not been successful. The Compiler and Publishers therefore offer their apologies, trusting that current copyright holders (if any) will understand that every endeavour has been made to trace them.

Acknowledgements

Contents

Glossary

Batik (*battek*, etc.): wax-resist dyeing process elevated to sublime intricacies in Java where the traditional hues are indigo and brown and the cloth is cotton.

Becak (*betja*, *becha*, etc.): a tricycle taxi in which passengers ride in front of the pedalling 'driver'.

Candi (*chandi*, *tjandi*): shrine or temple, as in Candi Borobudur.

Dukun: variously a herbalist, healer, diviner, and/or shaman.

Gamelan (*gamelang*, etc.): the Javanese percussion orchestra in which xylophones and gongs of multifarious pitches and tones predominate; inseparable from the *wayang*.

Kain: cloth or, commonly, a *sarung*, as in 'an ankle-length kain'.

Kampung (*kampong*, *campong*, etc.): a village or, within the city, a neighbourhood.

Kapok: cotton-like fibre from pods of the Kapok (*kapuk*) tree used for filling pillows, mattresses, and the like.

Kebaya: woman's blouse reaching below the waist and worn with a *sarung*, hence *sarung kebaya*.

Kraton: palace and administrative centre of Javanese rulers.

Keris (*kris*, *criss*, *creese*, etc.): double-edged daggers, valued as heirlooms and charms among the Javanese and other South-East Asians.

Pasar (*passer*, etc.): market-place, bazaar.

Passangrahan: a simple guest-house or inn provided by the government for touring officials and other travellers.

Prahu (*proa*, *prow*, *prau*, etc.): a boat; often used to indicate a traditional indigenous sailing craft or canoe.

Priangan (Preanger Districts): collectively the mountain provinces of west Java known for their cool climes, tea plantations, and panoramic scenery, including the famous Sundanese rice terraces.

Rijsttafel (*riz tavel*, etc.): rice table; colonial Dutch exaggeration of the multi-dish Indonesian feast.

Sarung (*sarong*): rectangular piece of cloth with the ends sewn together to form a tube, worn as a waist-to-ankle garment with a shirt or blouse by both men and women; also worn by women wrapped below the armpits when 'at home' or bathing in a river or other public place.

Saté: small pieces of marinated meat arranged on a bamboo skewer and roasted over a charcoal fire—a street-side favourite.

Sawah (*sawa*, etc.): wet rice, or padi (paddy) fields.

Waringen: the banyan tree.

Wayang (*wajang*): Javanese puppet theatre or dramatic performance; often used to connote *wayang kulit*, the buffalo-hide shadow puppets that embody the gods, demons, kings, warriors, princesses, and clowns of the *Ramayana–Mahabharata*-based legend cycle familiar to all Javanese.

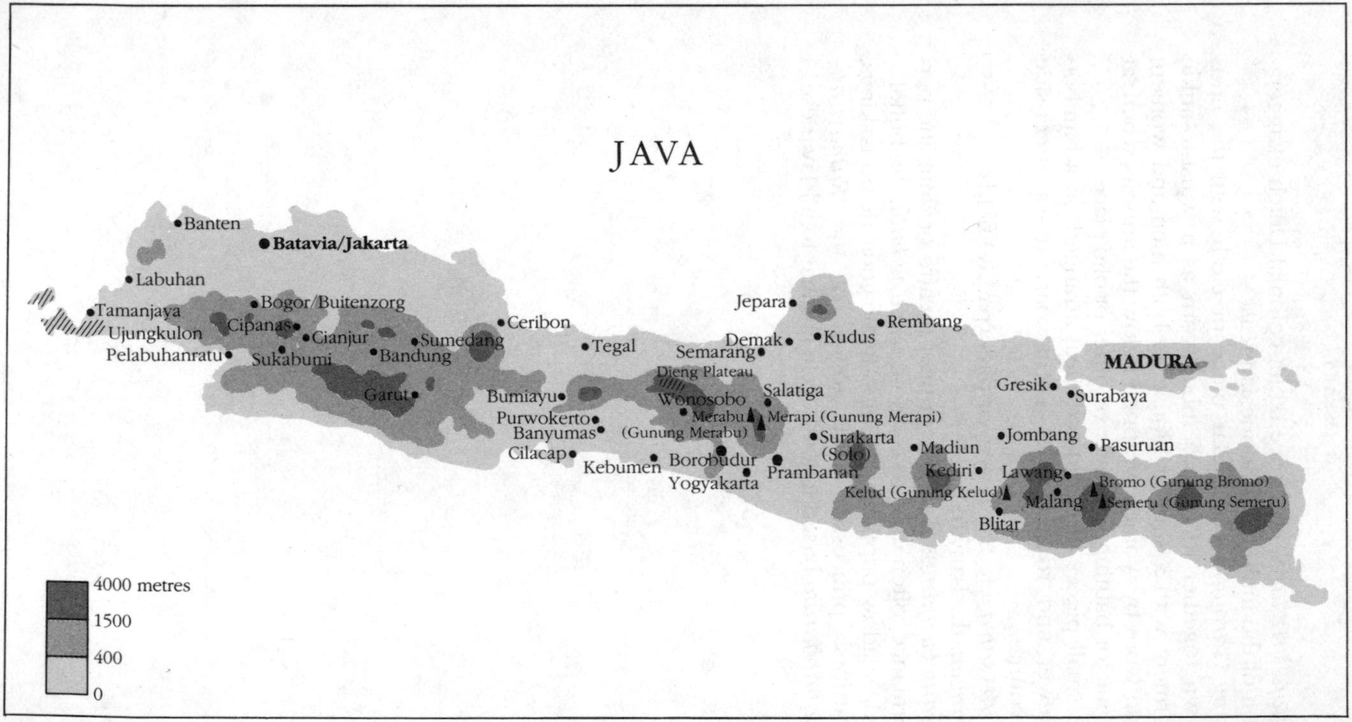

JAVA

Banten
Batavia/Jakarta
Labuhan
Tamanjaya
Ujungkulon
Pelabuhanratu
Bogor/Buitenzorg
Cipanas
Cianjur
Sukabumi
Bandung
Sumedang
Garut
Ceribon
Tegal
Bumiayu
Purwokerto
Banyumas
Cilacap
Kebumen
Dieng Plateau
Wonosobo
Merabu
(Gunung Merabu)
Borobudur
Yogyakarta
Semarang
Demak
Jepara
Kudus
Rembang
Salatiga
Merapi (Gunung Merapi)
Prambanan
Surakarta
(Solo)
Madiun
Kediri
Kelud (Gunung Kelud)
Blitar
Jombang
Lawang
Malang
Gresik
Surabaya
Pasuruan
Bromo (Gunung Bromo)
Semeru (Gunung Semeru)
MADURA
4000 metres
1500
400
0

Introduction

JAMES R. RUSH

> We have been out nearly one hundred days, and during this time travelled over nearly thirteen thousand miles, changed day into night, cold weather into hot, spring into autumn, one part of the world into another. We have changed from Americans to Africans, from Africans to South Africans, from South Africans to Africans again, and from Africans, pretty soon we shall be Javanese.[1]

THOMAS TURNER, passenger aboard his father's brig, the *Palestine*, set down these exuberant lines in 1832 as he made his way to the port of Batavia, capital of Java and of all Dutch possessions in the Indonesian Archipelago. The Turners were going to Java on business, to trade for rattan, sappanwood, and coffee. But young Thomas was obviously aware of something greater afoot. He was off to see the world, a traveller!

Although Thomas Turner's logbook of his journey to Java remains cloistered in the archives (where my sea-loving daughter, Margaret, recently discovered it), plenty of Java travellers' tales have seen the light of day in the years since Europeans first made their way there so many centuries ago. Excerpts from several of these are anthologized in this collection. I have arranged them chronologically so that they trace the encounter of Western travellers with Java from the years of 'discovery' through the very long period of Dutch rule there, and finally to Indonesian independence and nearly to the present. And, because this is a book about both Java *and* travel, as often as possible I have selected passages that show our travellers in motion.

Like Thomas Turner, the authors sampled here were outsiders to Java, most of them British or American. They came as explorers, merchants, scientists, journalists, and teachers, or simply *en touriste.* A few came, in the nineteenth century, to take a cure. One, Thomas Stamford Raffles, an Englishman, actually governed Java for four and a half years during the Age of Napoleon. But only one—Augusta de Wit, an Indies Dutchwoman who sometimes wrote in English—was truly at home there. Their writings fall broadly within the tradition of English-language travel writing, which began in the seventeenth century with accounts of the Grand Tour and voyages of discovery, climaxed in a great flood of books during the late nineteenth and early twentieth centuries, and survives today in the works of Paul Theroux, Bruce Chatwin, and others.

Although a few of the writers represented here developed a deep knowledge of Java, most gathered their facts along the way, some more cautiously than others. What they give us is nothing approaching a true Java, but rather their impressions of Java. These impressions are shaped by where the travellers happened to go, where they lodged, what they ate, who they spoke to, and for what reasons and with what foreknowledge they made their way to Java in the first place. Not surprisingly, a great many travellers to Java went to the same places, stayed in the same hotels, interacted with the same sorts of people, and, of course, ate the same sorts of food. Although until the nineteenth century travellers rarely penetrated beyond Batavia and the other coastal towns, by the second half of the nineteenth century, the traveller's Java had come inevitably to include the Botanical Gardens and Governor-General's mansion at Buitenzorg (now Bogor); the hill town of Bandung; the ruins at Prambanan and Borobudur; the principalities of Yogyakarta and Surakarta; a volcano crater or two, as well as tea, coffee, and sugar plantations. Each traveller in turn discovered the heat, the mosquitoes, the filthy canals of Batavia, and the incredible pastoral beauty of the west Java rice terraces. One after the other survived the same rites of passage: the Indies bath, or *mandi,* in which one ladles water over oneself from a large jar or basin; the Indies bolster, or Dutch Wife, an elongated pillow to separate the legs while sleeping on one's side—for coolness; the gargantuan Dutch colonial feast known as *rijsttafel*, and the shadow-puppet play that goes on and on and on the night long.

One of the pleasures of travel writing is that even accounts of places already familiar take on fresh allure and interest through the voices of individual travellers. It is a bit like listening to the same tune sung by a succession of performers, each with their own distinctive style and voice. Personal opinions and prejudices add spice to each traveller's tale, giving it the familiar tone of conversation or, better, gossip. A great many of the voices heard in this collection belong to British travellers. Because of their own vast empire, the British took a special interest in Dutch Java. As we shall see, their observations about it reflect both scorn and admiration for Dutch governance, and not a little envy.

The British Empire was already more vast than that of Holland's by many multiples in the 1860s when the naturalist Alfred Russel Wallace declared that Java 'may fairly be claimed to be the finest tropical island in the world'. By this he meant not only beautiful and wonderfully fertile and various in its natural life, but also a model for 'the moralist and the politician who want to solve the problem of how man may best be governed under new and varied circumstances'.[2] That Holland's Java should be the finest tropical island in the world was a proposition that subsequent British witnesses accepted with chagrin and that the Dutch smugly accepted as an accurate reflection of the facts. By the turn of the twentieth century most Western travel writers, including Americans who, from the 1890s, possessed several tropical islands of their own, simply repeated it. Eliza Scidmore, the American author of *Java: The Garden of the East* (1897), proclaimed Java 'ideal' and, quoting the famous Dutch novelist Multatuli, called it the priceless gem in 'that magnificent empire of Insulinde which winds about the equator like a garland of emeralds'.[3] Passages like this are robustly romantic and alluring, to be sure. But where Java travel writing is concerned, the key word is *empire*.

As a body of literature, Java travel writing, although in some ways distinctive as a genre, is closely related to, and indeed flows from, the same general body of experience, knowledge, and assumptions as do virtually all Western depictions of Java in the last several centuries. This vast body of material includes novels, stories, poems, newspapers, magazine articles, histories and other works of scholarship, the published works of government officials including laws and regulations, and any number of individually trivial artefacts of the times—songs, letters, snapshots. It is easier today to understand just how travel writing is nested within this larger web of words and images thanks to the pioneering insights of scholars such as Edward Said, who, in *Orientalism*, explained how the West's knowledge of the 'Orient' has been shaped by the nature of its relationship to the Orient, and Michael Adas, who, in *Machines as the Measure of Men*, unravels the interplay of experience and thought that led Western empire-builders to believe so confidently and unselfconsciously in their own natural superiority.[4] What Said and Adas teach us is that even a casual and open-minded Western traveller was inescapably implicated in the great enterprise of Empire. And, accordingly, that travel writing about Java is just as inescapably an integral part of the discourse of imperialism.

So, at one level, Java travel tales are truly about Batavia's hotels and famous *rijsttafel*, the Botanical Gardens of Buitenzorg, the pleasures and perils of travelling along the Great Post Road, the mountain tea districts and plains of ricefields and sugar plantations, about volcanoes, Borobudur, batik, and the quaint principalities with their proud pretenders, and about shadow-puppet plays and the other curious and pleasant ways of the Javanese and the island's other peoples. On another level, they are simultaneously about the advance of European power on the island, the growth of colonial institutions and infrastructure (railways, tea gardens, sugar mills), the reduction of the Javanese and other indigenous people to positions of subordination and economic servitude (rendering them 'natives' and their monuments 'antiquities'), and the evolution of a racist view of the world that, by the early twentieth century, was so thoroughly internalized among the white ruling caste that few of its members were fully aware of it, certainly not most of our travellers.

The first Western accounts of Java were a mix of the fanciful and true. The friar Odoric, who claimed to have visited Java in the early fourteenth century, noted that Java was completely inhabited, 'having great plenty of cloves, cubeb, and nutmegs, and all other kinds of spices'. The king, he said, was immensely rich and dwelled in a palace of gold and silver.[5] In the next century, Nicolo Conti saw feetless birds there—stuffed birds of paradise, perhaps—and accurately noted the people's love for gambling at cockfights.[6]

More detailed accounts emerged only in the early seventeenth century amidst the Northern European scramble for the spice trade and Asian markets. For quite a long time this advance of Holland and England into South-East Asia was a tentative undertaking, as the competing East India Companies (and others) vied for advantage, while almost wholly at the mercy of local people and circumstances. For these early travellers, Java was far away and dangerous. Englishman Edmund Scott, for example, found the low-lying harbour towns of Java 'very unwholesome' places that bred 'many diseases, especially among the strangers who resort thither'.[7] Strangers like himself. He catalogued the products of Bantam, the main trading city where both English and Dutch had established themselves, and noted the prevalence of fire and the providence of the local Chinese who 'plant, dress, and gather all the pepper, and sow the rice ... and absorb all the wealth of the land by their industry, from the indolent and idle Javanese'.[8] The king of Bantam, he says, however, was an 'absolute sovereign'. Trading at Bantam, fellow Englishman John Saris recorded, was very much under his thumb.[9] Less than twenty years later, Willem Ysbrantsz Bontekoe served on Dutch East India Company vessels in Asia. His *Memorable Description of the East Indian Voyage, 1618–25* was the first Dutch travel account of the eastern islands and a huge success in its day. Bontekoe tells of trading for cattle and fowls in the north Java town of Grise—where a Dutch merchant was already established—and of fetching pepper from Sumatra to Batavia, the newly established Dutch beachhead on the site of Jakarta. From islands offshore he hauls stones to complete its new fort, 'a pleasure to look at'.[10]

By the time the French Huguenot François Leguat and his companions made their voyage to the East Indies, a little less than a century later, Batavia had blossomed into a 'City and Suburbs ... inhabited by diverse Nations, viz. Dutch, French, Germans, Portugueses, Javans, Chineses and Moors'. Moreover, he writes, 'The Company is ... Absolute in this island'.

Java's 'petty sovereigns', as Leguat calls them, reign under the East India Company's protection and their territories are interpenetrated by Dutch 'forts and garrisons'.[11] Like many a male traveller, Leguat takes a special interest in Java's women, whom he describes as 'perfectly handsome' and 'extraordinarily amorous', a pleasant fact for Western men far from the company of Western women: 'The scarcity of that Sex oftentimes occasions the Christian Men to match with the Java woman.'[12] Java's colonial society was taking shape.

This was a society, as British traveller John Barrow witnessed it in the 1790s, in which the power and status of the constituent ethnic groups were more or less inversely proportional to their size. He describes the pecking order of a Batavian crowd. First come diverse Europeans, then Armenians, Persians, Arabs, and Indians, then the Chinese, then the Javanese ('loitering carelessly along, as if indifferent to everything around them'), the Malays, and finally Dutch Java's ubiquitous slaves 'from every nation and country of the East'. And at the very top: 'That class of men which bears a complete sway over the island is by much the least numerous; it is even rare to see a single ... right honourable high-born Dutchman condescending to walk the streets.'[13]

By this time, these 'right honourable high-born' Dutchmen of the East India Company have placed their stamp on the landscape. We learn from British traveller Joseph Banks, for example, that the islands off the coast of Java now bear the names Edam, Kuyper, and Onrust. And, as Banks also tells us, they have made Batavia 'by much the foremost town in the possession of Europeans in these parts'.[14] Indeed, the Frenchman Louis de Bougainville, who visited in the 1760s, claimed that the environs of Batavia 'adorned with houses and elegant gardens ... surpass those of the greatest cities in France, and approach the magnificence of those of Paris'.[15]

Somewhat oddly, travellers of the times were universal in describing Batavia as extremely unhealthy for Europeans. Several members of Banks's party succumbed to fevers during their stay, and he reported that of one hundred newcomers arriving from Europe it was rare for fifty to survive the first year.[16] Barrow commented on the 'pallid hue of the sickly European', and was horrified by mortality figures showing that the yearly death rate among the 'Dutch, half-cast and families' was four times higher than that of 'Natives and Malays'.[17] So why were Europeans enamoured of the place? The answer, clearly, is that for all its trouble Java was worth having. As de Bougainville noted, 'the luxury which prevails at Batavia is very striking'.[18] Taking a broader view, Banks, his own brush with death notwithstanding, concluded: 'Few parts of the world, I believe, are better furnished with the necessaries as well as the luxuries of life, than the island of Java.'[19] Java was a true colonial prize.

On the first day of the nineteenth century, this colonial prize slipped from the hands of the bankrupt Dutch East India Company into the uncertain grip of the Dutch government, which had itself, in 1795, fallen under the new empire of France. Java, too, in 1808 came under a Napoleonic administration, albeit one administered by an enthusiastic Dutch subaltern, Marshall Herman Willem Daendels. Then, in 1811, the British 'liberated' Java from the French usurpers and, under Thomas Stamford Raffles, governed it until 1816 when Holland recovered it.

These interregnums changed Java. For all their successes, the Dutch East India Company officials had for many years been culling its riches largely for private gain as the company itself slipped into ruin. They had governed Java lazily. Not so the new men. In true Napoleonic fashion, Daendels built a highway connecting the capital at Batavia to Bandung and central Java all the way to Surabaya. By making it easier to move soldiers, officials, and mail throughout the island, the new artery tightened the grip of the Batavia-based administration on the hinterland. At the same time, it opened Java as never before to travellers—and no nineteenth-century account is complete without an adventure along the Great Post Road. As Lieutenant-Governor for the British, Raffles advanced administrative reforms of a liberal stamp and threw himself into a comprehensive study of the island's population, industries, arts, and history. He initiated the restoration of the Buddhist monument Borobudur in central Java, henceforth to become an obligatory sight for all serious travellers. His encyclopaedic *History of Java* described in loving detail the brilliant civilization in decline that he earnestly hoped would remain under the enlightened stewardship of England. This was not to be, and for a time British travellers to Java wrote again of the 'noxious vapours' and 'dreadful maladies' of Batavia. Batavia itself, they observed, fell into decline following the establishment by Raffles of Britain's new free trade emporium at Singapore. Too bad. But as traveller George Windsor Earl commented after touring Java in the early 1830s, 'The Dutch are not the best colonists.'[20]

In thirty years' time, however, this view underwent a complete turn around. The bell-wether was a book by one James W. B. Money, a barrister of British India, titled *Java, or How to Manage a Colony* (1861). Money examined Java with an eye to solving the ills of India, recently beset by a frightening rebellion, and concluded that the Dutch colonial programme inaugurated in 1830—a system of forced production of export crops combined with infrastructure development and an intensification of the European administration known collectively as the Cultivation System—had transformed the island. The Cultivation System, he wrote, 'quadrupled the revenue, paid off the debt, changed the yearly deficit to a large yearly surplus, trebled the trade, improved the administration, diminished crime and litigation, gave peace, security, and affluence to the people, combined the interests of European and Native and, more wonderful still, nearly doubled an Oriental population, and gave contentment with the rule of their foreign conquerors to ten millions of a conquered Mussulman race'.[21] Never mind that Money got many of his facts wrong and that the system he praised was already under attack for its egregious abuses by the Dutch themselves. Money resurrected Raffles's image of a civilized but decadent Asian people living happily under enlightened white stewardship and, from this time until the 1940s, few Western travellers saw it otherwise.

Wallace, for example, toured Java for three and a half months in 1861 studying its flora and fauna and collecting specimens of birds and butterflies and nearly everything else. In his report of this trip, which forms Chapter 8 of his famous book, *The Malay Archipelago*, Wallace comments upon Java's great civilization of the past, noting that 'scattered through the country ... are found buried in lofty forests, temples, tombs, and statues of great beauty and grandeur'. Now, he says, 'a modern civilization of another type is ... spreading over the land. Good roads run through the country end to end;

European and native rulers work harmoniously together; and life and property are as well secured as in the best governed states of Europe.'[22] Here, in Java, the flaws of a 'semi-barbarous' people are abating under the influence of education and 'the gradual infusion of European blood'.[23] Wallace's precious plants and animals were the ostensible subject of his study, but he was almost equally interested in the natural order and progress of human society. To his trained scientist's eye, the 'improvement of the native population' was a work for Europeans.[24]

It is worth noting that as Wallace made his scientific observations and collected specimens in Java he dwelled in hotels and guest-houses operated strictly for Europeans, or as a guest of the colony's white residents; that he was attended everywhere by native assistants and servants, and that 'coolies' were always at hand to help with his baggage and equipment; that his movements everywhere were facilitated by Dutch officials and by the colony's good roads and other facilities, and that he was constantly in the orbit of colonial enterprises such as coffee and tea plantations and government projects like the Botanical Gardens at Buitenzorg. As intrepid as he was, in other words, Wallace's explorations were thoroughly shaped by Java's colonial structure and society. It could not have been otherwise; for Europeans, the land and the colony were one. This is why travel writing from the period is inescapably a literature of empire.

In this literature of empire, the lives and ways of the Javanese and other non-Europeans are often described with great colour, detail, amusement, and sometimes with a sort of condescending admiration; the Javanese especially are seen as childlike, simple, passive yet volatile, and clever. They are invariably subordinated to the real subject of this literature, which is the colony itself. And from the traveller's perspective, the colony was getting better all the time.

Henry O. Forbes, a naturalist who visited Java some twenty years after Wallace, reminds his readers of the old Batavia, 'a low-lying, close and stinking neighborhood'. 'All this is changed now,' he reports. 'Morning and evening, the train whirls in a few minutes the whole European population ... to and from the open salubrious suburbs, the new town, of fine be-gardened residences.'[25] In 1897, the American traveller

Eliza Ruhamah Scidmore reported that 'all Java is in a way as finished as little Holland itself, the whole island cultivated from edge to edge like a tulip garden, and connected throughout its length with post-roads smooth and perfect as park drives.... All the valleys, plains, and hillsides are planted in formal rows, hedged, trimmed, banked, drained, and carefully weeded as a flower bed.'[26] Scidmore is whisked from site to site by the colony's new trains and luxuriates in the company of Dutch planter families. (The hotel rooms, she complains, smell of mildew.)

Java—Dutch Java—is now completely safe. The dreaded diseases of yesteryear are increasingly kept at bay by modern health services. The dangers of travel along rough roads through the wilderness are a thing of the past. Moreover, the indigenous population poses no threat to travelling Europeans. The Natives, as they are always called, are thoroughly subordinated and, in the eyes of most travellers, happily so. The effect of all this for globe-trotting whites was most pleasant. Java, with its great natural beauty, its colourful and childlike population, its modern infrastructure, formed the perfect stage setting—a *tableau vivant*—both for the world's empire-builders and for the good times of the travellers themselves.

This view of Java pervades many travel accounts of the 1920s and 1930s and is most conspicuously rendered in the books of one Harriet W. Ponder, an Englishwoman. The Dutch, she wrote in *Java Pageant*, 'have made of Java ... the perfect colony'.[27] Moreover, 'despite all its perfection of roads, bridges, concrete and canals, despite its wealth of industries and electrical development, it yet manages to convey an impression that the whole thing is really a delightful entertainment, a sort of mystery play, dressed like the daintiest of light operas'.[28] In this light opera, Holland's costumed native subjects form a huge supporting cast for the main actors who are, of course, Europeans. And, as Wallace had remarked some seventy years before, the Dutch are doing nothing less than building a new civilization on the island. For in Java, Ponder notes, European settlement is as permanent as in Canada and Australia. And the innovations of the Dutch, 'things great and small ... have become part of the fabric of ordinary everyday modern colonial life and seem to us as permanent and inevitable as the sunshine and the ricefields'.[29] The future is personified not by Indonesian nationalists ('disturbing elements' among the Natives) but by 'fine healthy young Hollanders, sunburnt and sturdy and full of life, that you see on all sides as soon as you land in Java'.[30]

In a striking passage, Ponder captures the exhilaration of seeing Java from a speeding train, conveying at the same time metaphorically (and surely unconsciously) the delight for real and armchair travellers alike, of viewing the world's great colonial tableau from the special vantage point of the ruling race. The train moved along an embankment that 'seemed to have been designed for no other purpose than to provide the best of all possible observation points'. Thus elevated, she wrote: 'I had removed to another plane, whence, godlike and unassailable, I could ride at my ease and survey the ever changing scene as we raced along, high above crowded roads and villages, houses and gay gardens, and the chequer board of ricefields.'[31]

By the time Ponder completed her second book about Java—entitled *Javanese Panorama*—war had overtaken the colony, and Dutch Java as she had known it was no more. (She ends her preface with the words, 'May Java soon be returned to her former freedom, happiness, and prosperity'.)[32] The Japanese had wrested the island from the Dutch and held it for three and a half years. When they were defeated, Java's future was uncertain. The next round of travellers who recorded their journeys to Java encountered a different world. Peter Kemp, arriving in advance of the British armed forces at war's end, recalls his stay at the Hotel des Indes, once the queen of European hotels in the Far East: 'Gone were the enormous meals of *rijsttafel*, each requiring a posse of fifteen waiters to carry the dishes ...; and the high, airy suites with their wide verandas were stripped of most of their furniture and crowded with extra beds.' No longer embraced within a secure and dominant European élite, Kemp, like many travellers to come, felt 'bewildered ... and alone'.[33] For a time at least, Java was once again far away and dangerous.

To the Western traveller, Java during the Dutch years had become part of the great collective 'ours'. But with Indonesian independence—declared by some of Ponder's 'disturbing elements' in 1945 and achieved after four years of war and negotiations—it was no longer so. Java, along with the rest of Holland's giant tropical colony, had been reclaimed by Indonesians. It had drifted from 'our world' to 'their world', soon to be dubbed the Third World. Travellers, or so they wrote, suffered egregious inconveniences and discomforts; bureaucratic vexations of the sort familiar to visitors in Dutch Java were now worse as the newcomers were subjected to the wiles and whims of the island's new power holders, the former Natives. Some, like Harold Forster, who came to teach at Gadja Mada University, adjusted to the new winds with grace and hopefulness. Others, like Frank and Helen Schreider, traversed Java longing for the Indies of a certain 'old Dutchman' who had regaled them with nostalgic fantasies of the good old days. Still others, like American Arthur Goodfriend, went seeking ways to inoculate the Javanese against the infection of communism—that is, seeking ways to bring Java back into the fold. Others went simply for fun and experience and, like John Krich, author of *Music in Every Room: Around the World in a Bad Mood*, in hopes of turning a bad trip into a good story.

None of our travellers succeeded in 'becoming Javanese', as Thomas Turner of the *Palestine* put it, but in recent years a few have approached Java with the sincere hope of knowing the Javanese. Thus, Muslim Java is an important destination for V. S. Naipaul in his far-flung sojourn *Among the Believers*, and for Richard Critchfield who has visited and written about the same Javanese family and village since 1967.

We often associate travel with escape, I think, and hence with freedom. But we travellers are not so free. We go as the planes, trains, and buses go. Where we spend our days and nights is governed by timetables and the price of a ticket, a meal, a room. Moreover, none of us travels without baggage, and even the most earnestly dispassionate observers travel within a web of knowledge and ideas that determine what we look for and what we see. We have been told not to miss the Botanical Gardens—few of our travellers did—and to think twice before embarking on a long trip aboard a bus marked 'Deluxe'. (Sound advice, in my experience.) We cling to our guidebooks. Perhaps we have brought along an account of someone else's trip and read it as we go. So no traveller's tale is wholly original. Each one reflects the larger culture of the traveller's land of origin and of its relationship to the land of travel.

For most of us, however, this sobering awareness does not stop our longing for some kind of transcendent pleasure from travel. And in this sense 'Java' still beckons. For sheer exhilaration, it is hard to beat Ponder's epiphany aboard the colonial trains in the late 1930s, racing across Java's landscape on elevated rails 'godlike and unassailable'. For Ponder, speed and the heady elixir of racial superiority helped transform quotidian Java into that transcendent Java of the imagination. This Java is more elusive these days, judging from recent travel writing. Some who search, like the Schreiders, authors of *The Drums of Tonkin*, leave disappointed. But others, like Forster and Critchfield, find profound satisfactions in the new Java with its bright, optimistic students, and, in Critchfield's case, a friendship sustained over many visits. I have experienced these very pleasures in my own journeys to Java. And occasionally I have found that *other* Java too.

I think of my day—twelve years ago now—with Lesmana and Linda, exploring the terraces of Borobudur and the eloquent old stones of Prambanan. Driving back to Yogya at day's end, Lesmana turned off the main road and on to a country lane. Soon we approached a Hindu shrine, a temple of grey, mossy stones sunken as in a bowl in the ground. Submerged hundreds of years ago by volcanic ash, Lesmana told me, it was rediscovered only a few years ago, when a farmer plowing his ricefield struck the topmost stones. Now it had been excavated. Lesmana pulled over and we walked to the lip of the bowl. The headman of the local village wandered over. Some talk. Then we all sat quietly on the rim gazing at the ruin. The headman lit up a clove cigarette and Lesmana—who had studied at the Sorbonne—slipped a cassette of Parisian cabaret songs into the player in his jeep. The ancient temple, the scent of cloves, the incongruent music, the waning light, the peace of the surrounding ricefields—suddenly, miraculously, we were in that sweet other place travellers yearn for and roam the world to find. Ah, I thought, Java at last. Java.

1. Thomas Turner, 1832 journal aboard the brig *Palestine*, Collection 95, Vol. 14, Larkin Turner Papers, Manuscript Collection, G. W. Blunt White Library, Mystic, Connecticut.

2. Alfred Russel Wallace, *The Malay Archipelago* . . . , London: Macmillan and Company, 1869, p. 76.

3. Eliza Ruhamah Scidmore, *Java: The Garden of the East*, New York: Century Company, 1897, p. 334.

4. Michael Adas, *Machines as the Measure of Men*, Ithaca: Cornell University Press, 1989; Edward W. Said, *Orientalism*, New York: Vintage Books, 1978.

5. Robert Kerr (ed.), *General History and Collection of Voyages and Travels* . . . , Edinburgh: William Blackwood et al., 1811, Vol. 1, p. 408.

6. J. Winter Jones (trans.), 'The Travels of Nicolo Conti in the East in the Early Part of the Fifteenth Century', in R. H. Major (ed.), *India in the Fifteenth Century*, London: Hakluyt Society, 1857, p. 16.

7. Edmund Scott, 'Account of Java, and of the first Factory of the English at Bantam; with Occurrences there from the 11th February, 1602, to the 6th October, 1605', in Robert Kerr, *General History and Collection of Voyages and Travels* . . . , Edinburgh: William Blackwood et al., 1811, Vol. 8, p. 143.

8. Ibid., p. 144.

9. John Saris, 'Observations by Mr. John Saris of Occurrences during his abode at Bantam, from October, 1605 to October, 1609', in Robert Kerr, *General History and Collection of Voyages and Travels* . . . , Edinburgh: William Blackwood et al., 1811, Vol. 8, pp. 185–6.

10. Willem Ysbrantsz Bontekoe, *Memorable Description of the East Indian Voyage, 1618–25*, trans. C. B. Bodde-Hodgkinson and Pieter Geyl, London: George Routledge and Sons, 1929, p. 78.

11. François Leguat, *A Voyage to the East-Indies by Francis Leguat and His Companions*, London: R. Bonwicke et al., 1708, p. 189.

12. Ibid., p. 214.

13. John Barrow, *A Voyage to Cochinchina in the Years 1792 and 1793* . . . , London: T. Cadell and W. Davies, 1806, p. 203.

14. Joseph Banks, *Journal of the Right Hon. Sir Joseph Banks during Captain Cook's First Voyage in HMS Endeavour in 1768–71* . . . , London: Macmillan and Company, 1896, p. 381.

15. Lewis de Bougainville, *A Voyage Round the World*, trans. John Reinhold Forster, London, 1772, p. 426.

16. Banks, op. cit., pp. 371, 382.

17. Barrow, op. cit., pp. 202, 179.

18. Bougainville, op. cit., p. 428.

19. Banks, op. cit., p. 387.

20. George Windsor Earl, *The Eastern Seas, or Voyages and*

Adventures in the Indian Archipelago in 1832–33–34, London: Wm. H. Allen and Company, 1837, pp. 21, 23.

21. J. W. B. Money, *Java, or How to Manage a Colony*, London: Hurst and Blackett, 1861, Vol. 1, pp. viii, ix.

22. Wallace, op. cit., p. 76.

23. Ibid., p. 74.

24. Ibid., p. 86.

25. Henry O. Forbes, *A Naturalist's Wanderings in the Eastern Archipelago*, London: Sampson Low, Marston, Searle and Rivington, 1885, p. 8.

26. Scidmore, op. cit., p. 71.

27. Harriet W. Ponder, *Java Pageant*, London: Seeley Service and Company, 1935, p. 263.

28. Ibid., p. 17.

29. Harriet W. Ponder, *Javanese Panorama; More Impressions of the 1930s*, London: Seeley Service and Company, n.d. [*c*.1943], p. 258.

30. Ponder, *Java Pageant*, p. 250.

31. Ponder, *Javanese Panorama*, p. 223.

32. Ibid., p. 10.

33. Peter Kemp, *Alms for Oblivion*, London: Cassell, 1961, p. 84.

1
Friar Odoric Recalls Java, 1330

Odoric, the patron of long-distance travellers, was an Italian-born Franciscan friar who, in the early fourteenth century, plied the mission fields of southern Russia, India, and China and visited many places in between, including Java. In 1330, the year before he died, he dictated his account of these far-flung travels to the Italian friar William de Solona, and they subsequently became one of the most famous travel books of the times. Given the vagaries of Odoric's own memory and the temptation of later transcribers and compilers to embellish his text, it is hard to say just where fact turns to fancy in Odoric's brief description of Java. But cloves and nutmeg surely were plentiful in Java of the time—due to its proximity to the source of these precious spices in islands to the east—and in 1293, the Javanese histories tell us, a Javanese prince and future king, Kertarajasa Jayavarddhana, manoeuvred invading armies of Kublai Khan to his own advantage and then did indeed drive them away.

I then went to another island named Java, the coast of which is 3000 miles in circuit; and the king of Java has seven other kings under his supreme dominion. This is thought to be one of the largest islands in the world, and is thoroughly inhabited; having great plenty of cloves, cubebs, and nutmegs, and all other kinds of spices, and great abundance of provisions of all kinds, except wine. The king of Java has a large and sumptuous palace, the most lofty of any that I have seen, with broad and lofty stairs to ascend to the upper apartments, all the steps being alternately of gold and silver. The whole interior walls are lined with plates of beaten gold, on which the images of warriors are placed sculptured in gold, having each a golden coronet richly ornamented with precious stones. The roof of this palace is of pure gold, and all the lower rooms are paved with alternate square plates of gold and silver. The great khan, or emperor of Cathay, has had many wars with the king of Java, but has always been vanquished and beaten back.

'Travels of Oderic of Portenau, into China and the East, in 1318', in Robert Kerr (ed.), *General History and Collection of Voyages and Travel* ..., Edinburgh: William Blackwood et al., 1811, Vol. 1, pp. 408–9.

2
Nicolo Conti Finds Java beyond Civilization, 1444

Nicolo Conti learned Arabic in Damascus and from there made his way eastward to Baghdad, Persia, India, and finally to Java in twenty-five years of travel. He arrived back in Italy in 1444 and subsequently narrated an account of his adventures to the secretary of Pope Eugenius IV as penance for having renounced Christianity during his travels.

IN central India there are two islands towards the extreme confines of the world, both of which are called Java. One of these islands is three thousand miles in circumference, the other two thousand. Both are situated towards the east, and are distinguished from each other by the names of the Greater and the Less. These islands lay in his route to the ocean. They are distant from the continent one month's sail, and lie within one hundred miles of each other. He remained here for the space of nine months with his wife and children, who accompanied him in all his journeys.

The inhabitants of these islands are more inhuman and cruel than any other nation, and they eat mice, dogs, cats, and all other kinds of unclean animals. They exceed every other people in cruelty. They regard killing a man as a mere

Sixteenth-century map showing two Javas: Major and Minor. (Sebastian Münster, 'India Extrema XXIII Nova Tabula', *c.*1552)

jest, nor is any punishment allotted for such a deed. Debtors are given up to their creditors to be their slaves. But he who, rather than be a slave, prefers death, seizing a naked sword issues into the street and kills all he meets, until he is slain by some one more powerful than himself: then comes the creditor of the dead man and cites him by whom he was killed, demanding of him his debt, which he is constrained by the judges to satisfy.

If any one purchase a new scimiter or sword and wish to try it, he will thrust it into the breast of the first person he meets, neither is any punishment awarded for the death of that man. The passers by examine the wound, and praise the skill of the person who inflicted it if he thrust the weapon in direct. Every person may satisfy his desires by taking as many wives as he pleases.

The amusement most in vogue amongst them is cock-fighting. Several persons will produce their birds for fighting, each maintaining that his will be the conqueror. Those who are present to witness the sport make bets amongst themselves upon these combatants, and the cock that remains conqueror decides the winning bet.

'The Travels of Nicolo Conti in the East in the Early Part of the Fifteenth Century', translated from the original of Poggio Bracciolini by J. Walter Jones, in R. H. Major (ed.), *India in the Fifteenth Century* . . ., London: Hakluyt Society, 1857; reprinted New York: Burt Franklin, 1970, pp. 15–16. Reprinted with the permission of the Hakluyt Society.

3
Edmund Scott Describes Bantam, Java Major, 1606

From May 1603 to October 1605, Edmund Scott resided at the port city of Bantam in west Java as the agent for the English East India Company. Here he and a little band of English and Dutch—the vanguard of eventual European penetration and conquest—lived at the sufferance of the local king and faced the vagaries of fire and disease and their own intense rivalries. At the time, Java was merely a way station along the route to the richer Spice Islands of Ambon, Ternate, Tidore, and Banda. The establishment of a European-controlled enclave there was still several years off. Scott published this account of life in early seventeenth-century Java shortly after his return to England.

JAVA MAJOR is an island, which lyeth in 140 degrees of longitude, from the middle part of it, and in 9 degrees of latitude; being also about 146 leagues long, east and west, and some 90 broade, south and north. The middle part of which land is for the most part all mountaines, the which are not so steepe but that people doe travaile to the toppe of them, both on horsebacke and on foote. Some inhabitants dwell uppon those hils which stand next [i.e. nearest] to the sea; but in the verie middle of the land (so farre as ever I could learne) there is no inhabitants. But there are wild beasts of divers sorts, wherof some doe repaire neere the valleys adjoyning to the sea, and devoure many people. Towards the sea for the most part is low moorish [i.e. marshy] ground, wherein stand their principall townes of trade; the chiefest whereof lye on the north and north-east side of the island, as Chiringin [Ceribon], Bantam [Banten], Jackatra, and Jortan or Greesey [Gresik]. The which lowe ground is verie unholesome and breedeth many diseases, especially unto strangers which come thether; and yeeld no marchandise worthy trading for, or speaking of, but pepper; the which hath been brought in times past from all places of the land to Bantam, as the chiefe mart towne of that countrey. The which towne for trade doth farre exceede Achin or any towne or citie thereabouts. And pepper was wont to be brought thether from divers other countreys; which of late yeeres is not, by reason that the Dutchmen trade to every place to buy it up.

This towne of Bantam is about 3 miles in length; also very populous. There are three great markets kept in it every day, one in the forenoone and two in the afternoone. That especially which is kept in the forenoone doth so abound with people that they thronge together as in many faires in England. Yet I never saw any kinde of cattell to sell, by reason that there are verie few tame in the countrey. Their foode is altogether rise, with some hens and some fish, but in no great aboundance.

Bantam market in Scott's day, when people 'thronge together as in many faires in England'. (Engraving, Amsterdam, 1598)

The Javans houses are altogether built of great canes and some few small timbers, being sleight buildings. In many of the principall mens houses is good workemanship shewed, as carvings, &c. And some of the chiefest have a square brick rowme, being built in no better forme than a bricke-kill[n]; which is onely to put in all their houshold stuffe when fier commeth; but they seldome or never lodge nor eat in them.

There are many small rivers running thorough the towne. Also there is a good rhode for ships; whereby, if they were people of any reasonable capacitie, it would be made a verie goodly citie. Also it is walled round with a bricke wall, being verie warlike built, with flankers and turrets scowring everie way. I have been told by some that it was first built by the Chineses, and by others that it was first built by the Portingales; wherefore I cannot say certainely by which of them it was first built; but it is most likelye by the Chineses, by reason of the oldnesse of it, for in many places it is fallen to decay for want of repayring.

At the verie west end of this towne is the China towne; a narrow river parting them, which runneth crosse the end of the China towne up to the Kings court, and so through the middle of the great towne, and doth ebb and flowe, so that at a high water both galleys and junckes of great burthen may goe up to the middle of the great towne. This China towne is for the most part built of bricke; everie house square and flat overhead, having bordes and smale timbers or split canes layd over crosse, on which is layd bricks and sand, to defend them from fire. Over these bricke warehouses is set a shed, being built up with great canes and thatched; and some are built up with small timbers, but the greatest number with canes onely. Of late yeares, since wee came thether, many men of wealth have built their houses to the top all fire-free; of the which sort of houses, at our first comming, there was no more but the Sabindars house and the rich China marchants house; which, neverthelesse, by meanes of their windowes and sheds round about them, have been consumed with fire.

In this [China] towne stand the English and Dutch houses; which are built in the same manner, onely they are verie much bigger and higher than the ordinarie houses. And the Dutchmen of late, though with great cost and trouble, have built one of their houses up to the top all of bricke, fire-free, as they suppose.

The King of this place is absolute, and since the deposing and death of the late Emperour of Damacke [Demak] is held the principall king of that island. He useth alway marshall law uppon any offender whome hee is disposed to punish. More, if any private mans wife, or wives, bee taken with dishonestie (so that they have good proofe of it), they have power in their owne hands to cause them presently to be put to death, both man and woman. And for their slaves, they may execute them for any small fault. If the King send for any subject or stranger dwelling or being in his dominions, if he send a man, the partie may refuse to come; but if once he send a woman, hee may not refuse nor make no excuse. Moreover, if any inferiour bodie have a suit to a man of authoritie, if they come not themselves, they alwayes send a woman; neither doe they ever come or send but they present the party they sue too with some present, be their suite never so small. To everie wife that a Javan (being a free man) marrieth, he must keep 10 women slaves, which they as ordinarie use as their wives; and some of them keepe for every wife 40 slaves, for, so they keepe 10, they may have as many more as they will; but they may have but 3 wives onely.

The Javans are generally exceeding proud; although extreame poore, by reason that not one amongst a hundreth of them will worke. The gentlemen of this land are brought to be poore by the number of slaves that they keepe, which eat faster than their pepper or rise groweth. The Chineses do both plant, dresse and gather the pepper, and also sowe their rise; living as slaves under them, but they sucke away all the wealth of the land, by reason that the Javans are so idle. And a Javan is so proude that he will not endure one to sit an inch higher in height above him, if hee bee but of the like calling. They are a people that do very much thirst after blood. If any Javan have committed a fact worthy of death and that he be pursued by any, whereby he thinketh hee shall die, he will presently draw his weapon and cry *Amucke*, which is as much [as] to say: I am resolved; not sparing to murther either man, woman, or childe which they can possibly come at; and he that killeth most dieth with greatest honor and credit. They will seldom fight face to face with one another, or with any other nation, but do altogether seek revenge of their enemie cowardly, albeit they are, for the most part, men of a goodlie stature. Their law for murther is to pay a fine to the King, and that but a small summe; but evermore the friends of the partie murthered will be revenged on the murtherer or his kindred. So that the more they kill one another, the more fines or profite hath their King.

Their ordinarie weapon which they weare is called a crise. It is about two foote in length; the blade beeing waved and crooked too and fro ... and withall exceeding sharpe; most of them having the temper of their mettall poysoned, so that not one amongst five hundred that is wounded with them in the bodie escapeth with his life. The handles of these weapons are either of horne or wood, curiously carved in the likenesse of a divell, the which many of them do worship. In their warres their fight is altogether with pikes, dartes, and targets. Of late some few of them have learned to use their peeces [i.e. muskets], but verie untowardly.

The gentilitie, both men and women, never goe abroad but they have a pike borne before them. The apparrell of the better sort is a tucke on their heads, and about their loynes a faire pintado [printed or painted cloth]; all the rest of their bodies naked. Sometimes they will weare a loose coate, somewhat like a mandillion [a loose overcoat], of velvet, chamlet cloth, or some other kind of silke; but it is but seldome, and uppon some extraordinarie ocasion....

The men, for the most part, have verie thicke curled haire on their heads; in which they take great pride, and often will goe bareheaded to shew it. The women goe all bareheaded; some of them having their haire tucked up like a carthorse tayle, but the better sort doe tucke it up like our riding geldings tayles. About their loynes they weare of the same stuffes which I have before mentioned; alwayes having a faire girdle or pintado of their countrey fashion throwne over one of their shoulders, which hangeth downe loose behinde them.

The principallest of them are most religious; but they very seldome goe to church. They doe acknowledge Christ to be a great prophet, whom they call *Naby Isat* [Nabi Isa, the prophet Jesus]; and some of them do keepe of Mahomets priestes in their houses. But the common people have very little knowledge in any religion; onely they say there is a God which made heaven and earth and them also. Hee is good (they say) and will not hurt them; but the Divell is naught [i.e. bad] and will doe them hurt; wherefore many of them, for want of knowledge, doe pray to him onely, for feare least he should hurt them. And surely, if there were men of learning (which were perfect in their language) to instruct them, a number of them would be drawen to the true fayth of Christ, and also would be brought to civilitie. For many which I have reasoned with concerning the lawes of Christians have liked all well, excepting onely their pluralitie of women, for they are all very lasciviously given, both men and women.

The better sort which are in authoritie are great takers of bribes; and all the Javans in generall are badd paymasters when they are trusted. Notwithstanding, their lawes for debts are so strict that the creditour may take his debtour, his wives, children, and slaves, and all that hee hath, and sell them for his debt....

They delight much in ease and musicke. And for the most part they spend the day sitting crosse-legged like a taylor, whitling of stickes; whereby many of them become very good carvers to carve their cryse handles; and that is all the worke the most of them indevour to doe.

They are very great eaters; but the gentlemen allow their slaves nothing but rice, sodden in water, with some rootes and hearbes. And they have a certaine hearbe called *bettaile* [betel], which they usually have carryed with them wheresoever they goe, in boxes or wrapped up in cloath, like a sugerloafe; and also a nutt called *pinange* [areca-nut]; which are both in operation very hott, and they eate them continually, to warme them within and keepe them from the fluxe. They doe likwise take much tobacco and opium.

Edmund Scott, 'The Description of Java Major, and the Manner and Fashions of the People, both Javans and Chineses, Which Doe There Inhabit', in William Foster (ed.), *The Voyage of Sir Henry Middleton to the Moluccas, 1604–1606*, London: Hakluyt Society, 1943, pp. 168–73; Edmund Scott's *An Exact Discourse of the Subtilties, Fashishions, Pollicies, Religion, and Ceremonies of the East Indians*... originally published London: Walter Burre, 1606. Reprinted with the permission of the Hakluyt Society.

4
François Leguat Reveals the Strange Ways of Batavia, 1708

In 1619, the Dutch East India Company established a fortified enclave at Jakatra on Java's north coast. They named it Batavia. Batavia became the hub of the company's expanding interests in the region and the eventual capital of all Netherlands India, or the Dutch East Indies. Western travellers to Java invariably disembarked at Batavia, and most accounts by Java travellers include observations about the city—on balance, highly unfavourable. François Leguat was an involuntary visitor. He and a small band of fellow Huguenots had run afoul of the Dutch governor of Mauritius, who arrested them and, following an escape attempt, had them transported to Batavia in 1698. There they languished in custody for two more years. Leguat eventually made his way back to Europe and settled in England. In 1708, he published an account of his adventures and misadventures. In these passages he comments on the new Dutch capital, local customs, and women.

BATAVIA including the City and Suburbs, is inhabited by divers Nations, viz. Dutch, French, Germans, Portugueses, Javans, Chineses and Moors. The Languages most in use are, Dutch, Malay, Portuguese and Chinese.

The Company is as it were Absolute in this Island; a great number of petty Sovereigns reigning there under their Protection: Nay, the Emperor of Japar [Jepara], who is by far the most Potent of any of them, cannot be said to be entire Sovereign of his Country, since the Hollanders have divers Forts and Garrisons in it. . . .

Every one knows what the Betel-Leaves, and Arequa Nuts are, which all the Natives of this Island, both Men, Women and Children chaw incessantly to fortifie their Gums and Stomach, for sometimes they swallow the Juice. This Juice is as red as Blood, and gives a like Tincture to the Spittle, which it provokes abundantly, so that all that use this Drink, have their Lips continually bloody as it were, which is no pleasant sight to look upon. When you are not accustom'd to this Drug, you find its Tast insupportably sharp, but otherwise it

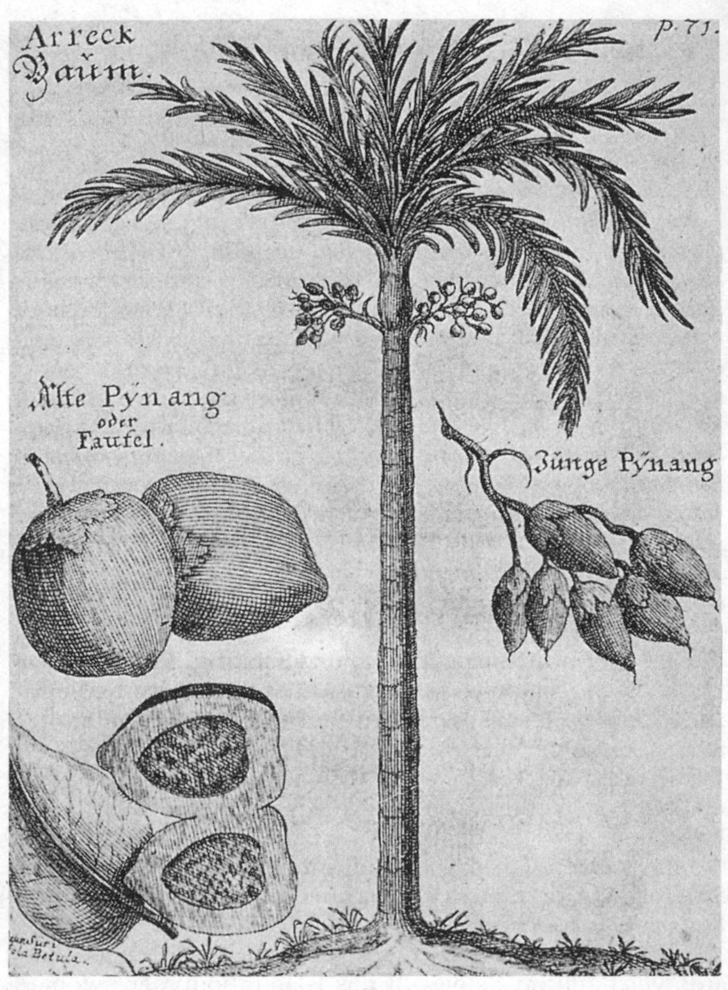

The areca-nut which, with betel leaves, 'all the Natives of this island ... chaw incessantly to fortifie their Gums and Stomach....' (From Georg Meister, *Der Orientallis Indische Kunst-und Lust-gartner*, Dresden, 1692)

becomes like Tobacco, and you find it difficult to leave it. If this Betel strengthens the Gums, as all say it does, with all my Heart, but I'm sure at the same time it blackens the Teeth in that frightful manner, that these People must needs be ignorant of the sweetness and charms of a fair Mouth. Betel is a Shrub, shap'd somewhat like a Pepper-Tree, but it has triangular Leaves, and is green all the year round. The Tree that bears the Nut call'd Arequa, is very tall and straight. They commonly wrap up a quarter of an Arequa-Nut in some Betel-Leaves, and so chaw them together: Some add a little Slack'd Lime, but that is not in use at Batavia. . . .

The Sport of Cock-fighting . . . is one of the greatest and most common Diversions of this Island. They breed up great numbers of these Animals on purpose; and arm them with sharp Iron Spurs, which they make use of with greater Dexterity than Force. The Javans are the Managers of these Sports, and whoever will, may come to them Gratis: Almost every Body is concern'd in Wagering more or less, and sometimes considerable Sums are lay'd. Whereas in England, where this Diversion is likewise common, they disfigure their Cocks by cutting off their Tails, and plucking out Feathers out of other parts of their Body, they here leave them in their natural State. 'Tis true, they are not so nimble as the English Cocks, but that Inconvenience being equal on both sides, it is no advantage to either, and the Combatants appear Nobler, and more fierce. Some of these Cocks have greatly enrich'd their Masters. . . .

The Javan Women, according to common Report, are extraordinary amorous, and what is uncommon, their Passion is no less constant than strong. They frequently make use of Philtres, which I have been assur'd they administer to their Lovers with Success: And when they suspect that any of them have been faithless to them, they do not fail to regale them with such a Drug, as quickly sends them to the Kingdom of Moles, so that it requires a Mans Consideration twice, before he engages with those sorts of Females once. There are a great many who not being so much expos'd to fatigues as the Men, are not near so Tawny, and who might pass even in Europe for Genteel. They have likewise good Faces, especially the younger sort, according to the notion we have of Beauty. Their pretty swelling Breasts have no conformity with the dangling Duggs of the homely Africans near the Cape. Their Complexions are fine and good, tho' a little brownish, their Hands fair, their Air soft, their Eyes sprightly, and their Laughing agreeable: To put all together, there are many of them that are perfectly handsom. I have seen some Dance the most charmingly that could be. They go about Streets with a sort of Drummer after their mode, who beats time to their Motions; and after the young Wench has ended her Dance, one gives her something for Encouragement. Another Charm they have, is, that they are extreamly neat and cleanly, their Religion obliging them to wash themselves all over several times a Day, and their Custom being, as I have already observ'd, to clear their Skins of all that hinders them from being extreamly soft and smooth. After all this, I know not whether Java may not pass for a gallant Country.

But after I have commended these jolly Women so much, I can't help adding to their Disadvantage, that if all I have heard be true, they are not extraordinary faithful to their Husbands; nevertheless, they seem extreamly Submissive. They lie groveling along upon the Ground while their Husbands sit, especially if there be any Strangers there (for it may be inferr'd from all I have said, they do not conceal themselves like the Chinese, or other Mahometans of Turkey, Persia, and elsewhere). But generally speaking, these couchant Postures signifie little to their Honesty, and are of no more Signification, than *your most humble Servant* at the end of a Letter.

These Women go in their Hair, and have for Habit a short Wastcoat with little Sleeves lac'd before, which fits close without joining, and being cut sloping a-top, discovers great part of their Breasts. Under this Corselet which hardly reaches to their Hipps, they wrap their Bodies in a sort of Scarf of divers Colours, which serves them for a Petticoat, and which being light and thin, requires two or three folds to keep them warm. This covers them to their Ancles, but as they wear no Smocks, there is always a List of brownish Flesh (which it may be, would not be better if it were altogether white) seen between the bottom of the Wastcoat, and the upper part of the Scarf. This covering sitting close to their Bodies, displays the bad shapes of some of them, as it does the good of others, which last has, I know not what bewitching effect. The richest of Women wear Slippers, which perhaps may be a mark of Distinction, because few wear them, tho' they cost but little.

When these Women marry any Hollanders, or other Christians, they are likewise oblig'd to espouse the Christian Religion. God only knows what sort of Christianity that is, for from the third and fourth Generation, the Children that are born of these Marriages, always go after the Javan Mode. It is these sorts of Converts that generally fill the Malay Church before mention'd: The number of Converts is much less, in that the Men have not the same Motives for Conversion. The Christian Women being but very few for the Christians themselves, the Javans, can get none of them, let them be as much Converts as they please, whereas the scarcity of that Sex often-times occasions the Christian Men to match with the Javan Women.

François Leguat, *A Voyage to the East-Indies by Francis Leguat and His Companions*, London: R. Bonwicke et al., 1708, pp. 189, 183, 186, 212–14.

5

Alexander Hamilton Reconnoitres Java's North Coast with Commerce in Mind, 1727

Englishman Alexander Hamilton spent the years 1688 to 1723 'trading and travelling by sea and land, to most of the countries and islands of commerce and navigation between the Cape of Good-hope and the island of Japon'—as the title-page of his New Account of the East Indies *of 1727 proclaims. Passing along the north coast of Java, he comments on the products of Bantam, Batavia, Semarang, Japara, Rembang, Surabaya, and several other towns and takes note of the advancing presence of the Dutch. Travellers, like the Dutch themselves, were still confined to the apron of Java.*

THE first Place of Commerce on the West End of Java, is the famous Bantam, where the English and Danes had their Factories flourishing till *Anno* 1682, at which Time the neighbourly Dutch fomented a War between the old King of Bantam and his Son, and because the Father would not come into their Measures, and be their humble Slave, they struck in with the Son, who was more covetous of a Crown than of Wisdom. They, with the Assistance of other Rebels, put the Son on the Throne, and took the old King Prisoner, and sent him to Batavia; and, in 1683, they pretended a Power from the new King to send the English and Danes a packing, which they did with a great Deal of Insolence, according to Custom. They next fortified, by building a strong Fort within a Pistol-shot of one that the old King had built before to bridle their Insolence.

The only Product of Bantam is Pepper, wherein it abounds so much, that they can export 10000 Tuns per Annum. The Road is good, and secure for the Safety of Shipping. It is in a pleasant Bay, wherein are several small Islands, which retain their English Names still; and the Natives still lament the Loss of the English Trade among them, but the King has much more Reason than his Subjects to regret the Loss of their Commerce. The Good-will the Natives bear to the Dutch may be conjectured from their Treatment, when they find an Opportunity, for if an Hollander goes but a Musket-shot from their Fort, it is five to one if ever he returns, for they are dextrous in throwing a Lance, or shooting of poisoned Darts thro' a wooden Pipe or Trunk; and the King never redresses them, pretending the Criminal cannot be found.

Batavia is about 20 Leagues to the Eastward of Bantam, and a great Number of small Islands ly scattered in the Way, too tedious to mention. Pullo-panjang off Bantam, and Edam off Batavia are the most conspicuous, and the Road of Batavia is almost surrounded with Islands, some of them inhabited, and some not. Its Topography I'll refer to another Time, with some historical Accounts of it, both ancient and modern.

Cheraboan [Ceribon] is the next Colony on the Coast, to the Eastward of Batavia, belonging to the Dutch, where they have a Fort and a small Garison.

Tagal [Tegal] is also a Dutch Settlement, with a small Fort for its Defence; and there is no other remarkable Place till we come to Samarang [Semarang], a good Colony, with a Fort of Mud and Wood to defend it. Damack [Demak] and Coutus [Kudus], two Places that ly between Samarang and Japara, are noted, one for the Abundance of Rice that it exports, and the other for great Quantities of good Sugars that it produces. They are peopled mostly with Chinese, and so is Japara, which formerly had an English Factory, but now is altogether in the Dutch Hands. It is defended by two Forts, one on an Hill, and the other in a Plain, where the Town stands, and has a small River to wash its Walls. The Road is secured by two Islands that ly about a League off the Town. I bought good white Sugar in Cakes here for two Dutch Dollars per Pecul, being 140 lb English suttle Weight.

Tampeira [?] is the next Place to the Eastward, and to the Eastward of it is Rambang [Rembang], about 2 Leagues from it, where the Dutch have a small wooden Fort, and a little Garison of sixteen Men. Those two afford nothing but excellent Teak Timber for building. And to the Eastward of Rambang is Sorobay [Surabaya], which lies within the Island Madura, and, I believe, is the eastmost Settlement the Dutch have on the Island of Java. It produces much Pepper, some Bees-wax and Iron. Sorobay is about 125 Leagues to the Eastward of Batavia, and the Country, along Shore, as pleasant and fruitful in Grain and Fruits as any in the World. Tame Cattle and wild Game are very plentiful, good and Cheap.

Alexander Hamilton, *A New Account of the East Indies* ..., Edinburgh: John Mosman, 1727; reprinted William Foster (ed.), London: The Argonaut Press, 1930, pp. 68–9.

6
Sir Joseph Banks Survives Three Months in Java, 1770

When the HMS Endeavour *put in at Batavia in September 1770, its passengers were 'rosy and plump', writes Joseph Banks, who accompanied Captain James Cook on his round-the-world voyage. Three months later, several of their number were dead and others, like Banks himself, had barely escaped death. Batavia was a very unhealthy place for travellers. Java itself, however, as Banks tells it, was 'uncommonly fruitful' and exceptionally 'furnished with the necessaries as well as the luxuries of life'.*[1] *In the following passages, he describes how he and his companions faced the plague and gives his observations about the various people who inhabited the city, including numerous slaves. (Owning slaves remained legal in the colony until 1860.)*

9th [October 1770.]

BEFORE four we were at anchor in Batavia road. A boat came immediately on board us from a ship which had a broad pendant flying; the officer on board inquired who we were, etc., and immediately returned. Both he and his people were pale almost as spectres, no good omen of the healthiness of the country we had arrived at. Our people, however, who might truly be called rosy and plump (for we had not a sick man among us), jeered and flaunted much at their brother seamen's white faces....

[*10th*.] Ever since our first arrival here we had been universally told of the extreme unwholesomeness of the place, which we, they said, should severely feel on account of the freshness and healthiness of our countenances. This threat, however, we did not much regard, thinking ourselves too well seasoned to variety of climates to fear any, and trusting more than all to an invariable temperance in everything, which we had as yet unalterably kept during our whole residence in the warm latitudes. Before the end of the month, however, we were made sensible of our mistake....

21st. We now began sensibly to feel the ill effects of the unwholesome climate we were in. Our appetites and spirits were gone, but none were yet really sick except poor Tupia and Tayeto, both of whom grew worse and worse daily, so that I began once more to despair of poor Tupia's life. At last he desired to be moved to the ship, where, he said, he should breathe a freer air clear of the numerous houses, which he believed to be the cause of his disease, by stopping the free draught. Accordingly on the 28th I went down with him to Kuyper, and on his liking the shore had a tent pitched

[1]Joseph Banks, *Journal of the Right Hon. Sir Joseph Banks during Captain Cook's First Voyage in HMS Endeavour in 1768–71* ..., London: Macmillan and Company, 1896, p. 387.

for him in a place he chose, where both sea and land breezes blew right over him, a situation in which he expressed great satisfaction.

The seamen now fell ill fast, so that the tents ashore were always full of sick. After a stay of two days I left Tupia well satisfied in mind, but not at all better in body, and returned to town, where I was immediately seized with a tertian, the fits of which were so violent as to deprive me entirely of my senses, and leave me so weak as scarcely to be able to crawl downstairs. . . .

6th [November]. In the afternoon of this day poor Mr Monkhouse departed, the first sacrifice to the climate, and the next day was buried. Dr Solander attended his funeral, and I should certainly have done the same, had I not been confined to my bed by my fever. Our case now became melancholy, neither of my servants were able to help me, no more than I was them, and the Malay slaves, whom alone we depended on, naturally the worst attendants in nature, were rendered less careful by our incapacity to scold them on account of our ignorance of the language. When we became so sick that we could not help ourselves, they would get out of call, so that we were obliged to remain still until able to get up and go in search of them.

9th. This day we received the disagreeable news of the death of Tayeto. . . .

10th. Dr Solander and I still grew worse and worse, and the physician who attended us declared that the country air was necessary for our recovery; so we began to look out for a country house, though with a heavy heart, as we knew that we must there commit ourselves entirely to the care of the Malays, whose behaviour to sick people we had all the reason in the world to find fault with. For this reason we resolved to buy each of us a Malay woman to nurse us, hoping that the tenderness of the sex would prevail even here, which indeed we found it to do, for they turned out by no means bad nurses. . . .

24th [November]. Dr Solander had recovered enough to be able to walk about the house, but gathered strength very slowly. I myself was given to understand that curing my ague was of very little consequence while the cause remained in the badness of the air. The physician, however, bled me, and gave me frequent gentle purges, which he told me would make the attacks less violent, as was really the case....

7th. We received the agreeable news of the ship's arrival in the road, having completed all her rigging, etc., and having now nothing to take in but provisions and a little water. The people on board, however, were extremely sickly, and several had died, a circumstance necessarily productive of delays; indeed, had they been strong and healthy we should have been before now at sea.

Dr Solander had changed much for the better within these two last days, so that our fears of losing him were entirely dissipated, for which much praise is due to his ingenious physician, Dr Jaggi, who at this juncture especially was indefatigable.

16th. Our departure being now very soon to take place, I thought it would be very convenient to cure the ague, which had now been my constant companion for many weeks. Accordingly I took decoction of bark plentifully, and in three or four days missed it. I then went to town, settled all my affairs, and remained impatient to have the day fixed.

24th. The 25th, Christmas Day by our account, being fixed for sailing, we this morning hired a large country proa, which came up to the door and took in Dr Solander, now tolerably recovered, and carried him on board the ship, where in the evening we all joined him....

[*Peoples of Batavia*.]

The town of Batavia, though the capital of the Dutch dominions in India, is so far from being peopled with Dutchmen, that I may safely affirm that of the Europeans inhabiting it and its neighbourhood, not one-fifth part are Dutch. Besides them are Portuguese, Indians and Chinese, the two last many times exceeding the Europeans in number....

The Dutch, Portuguese, and Indians here are entirely waited upon by slaves, whom they purchase from Sumatra, Malacca, and almost all their eastern islands. The natives of Java only have an exemption from slavery, enforced by strong penal laws, which, I believe, are very seldom broken. The price of these slaves is from ten to twenty pounds sterling apiece; excepting young girls, who are sold on account of their beauty; these sometimes go as high as a hundred, but I

Multiethnic Batavians along the canal. (From Johan Nieuhof, *Gedenkwaerdige Zee en Lantreize door de voornaemste Landschappen van West en Oost-Indien*, Amsterdam, 1682)

believe never higher. They are a most lazy set of people, but contented with a little; boiled rice, with a little of the cheapest fish, is the food which they prefer to all others. They differ immensely in form of body, disposition, and consequently in value, according to the countries they come from.... The island of Bali sends the most honest and faithful, consequently the dearest slaves, and Nias, a small island on the coast of Sumatra, the handsomest women, but of tender, delicate constitutions, ill able to bear the unwholesome climate of Batavia. Besides these are many more sorts, whose names and qualifications I have entirely forgotten....

Extraordinary as it may seem, there are very few Javans, that is descendants of the original inhabitants of Java, who live in the neighbourhood of Batavia, but there are as many sorts of Indians as there are countries the Dutch import slaves from; either slaves made free or descendants of such. They are all called by the name of *oran slam*, or *Isalam*, a name by which they distinguish themselves from all other religions, the term signifying believers of the true faith. They are again subdivided into innumerable divisions, the people from each country keeping themselves in some degree distinct from the rest. The dispositions generally observed in the slaves are, however, visible in the freemen, who completely inherit the different vices or virtues of their respective countries.

Many of these employ themselves in cultivating gardens, and in selling fruit and flowers; all the betel and areca, called here *siri* and *pinang*, of which an immense quantity is chewed by Portuguese, Chinese, Slams, slaves, and freemen, is grown by them. The lime that they use here is, however, slaked, by which means their teeth are not eaten up in the same manner as those of the people of Savu who use it unslaked. They mix it also with a substance called *gambir*, which is brought from the continent of India, and the better sort of women use with their chew many sorts of perfumes, as cardamoms, etc., to give the breath an agreeable smell. Many also get a livelihood by fishing and carrying goods upon the water, etc. Some, however, there are who are very rich and live splendidly in their own way, which consists almost entirely in possessing a number of slaves.

In the article of food no people can be more abstemious than they are. Boiled rice is of rich, as well as of poor, the principal part of their subsistence: this with a small proportion of fish, buffalo or fowl, and sometimes dried fish and dry shrimps, brought here from China, is their chief food. Everything, however, must be highly seasoned with cayenne pepper. They have also many pastry dishes made of rice flour and other things I am totally ignorant of, which are very pleasant: fruit also they eat much of, especially plantains.

Their feasts are plentiful, and in their way magnificent, though they consist more of show than meat: artificial flowers, etc., are in profusion, and meat plentiful, though there is no great variety of dishes. Their religion of Mahometanism denies them the use of strong liquors: nor do I believe that they trespass much in that way, having always tobacco, betel, and opium wherewith to intoxicate themselves....

The language spoken among them is entirely Malay, or at least so called, for I believe it is a most corrupt dialect. Notwithstanding that Java has two or three languages, and almost every little island besides its own, distinct from the rest, yet none use, or I believe remember, their own language, so that this *Lingua Franca* Malay is the only one spoken in this neighbourhood, and, I have been told, over a very large part of the East Indies.

Joseph Banks, *Journal of the Right Hon. Sir Joseph Banks during Captain Cook's First Voyage in HMS Endeavour in 1768–71 ...*, London: Macmillan and Company, 1896, pp. 366, 369, 371–2, 375–6, 401, 404–7.

7
Sir John Barrow Visits Dutch Batavia, 1792

John Barrow, Secretary to the British Admiralty and chronicler of travels with Lord Macartney to China and South Africa, stopped in Batavia during his voyage to Cochin-China of 1792–3. In the passages below, he comments on the city and its multiethnic population (penning, perhaps, the literary forebear of Noel Coward's famous line about 'mad dogs and

Englishmen'), takes us feasting with a high Dutch official, and introduces us to Batavia's notorious 'nine o'clock flower'.

IN making choice of the present site of the city of Batavia, the predilection of the Dutch for a low swampy situation evidently got the better of their prudence; and the fatal consequences that have invariably attended this choice, from its first establishment to the present period, irrefragably demonstrated by the many thousands who have fallen a sacrifice to it, have nevertheless been hitherto unavailing to induce the government either altogether to abandon the spot for another more healthy, or to remove the local and immediate causes of a more than ordinary mortality. Never were national prejudices and national taste so injudiciously misapplied, as in the attempt to assimilate those of Holland to the climate and the soil of Batavia. Yet such has been the aim of the settlers, which they have endeavoured to accomplish with indefatigable industry. An extended plain of rich alluvious land; with a copious river serpentizing through it, in a stream of so easy and gentle a current that the water with great facility was capable of being conducted at pleasure; a tract of country holding out such easy means of being intersected by canals and ditches, and embellished with fish ponds; of being converted into gardens and villas, where draw-bridges for ornament and *trek-schuyts* [tow-boats] for pleasure and convenience could be adopted, presented temptations too strong for Dutch taste to resist. Nothing, however, can possibly be more gratifying to the eye than the general appearance of the country which surrounds Batavia. Here no aridity, no sterility, no nakedness even partially intervene between the plantations of coffee, sugar, pepper, rice, and other valuable products, which are enclosed and divided by trees of the choicest fruits. In the immediate vicinity of the city, the extensive gardens of the Dutch, embellished with villas in the Oriental style, furnished with every convenience that a luxurious and voluptuous taste can suggest, are charming to behold from a little distance, but do not improve by a nearer acquaintance. The vitiated taste of Holland, delighting in straight avenues, trimmed hedges, myrtles and other evergreens cut into the *walls of Troy*, and flower-beds laid out in circles, squares, and polygons, are no less offensive to the eye than the numerous ditches and fish-ponds, from their stench and exhalations, are injurious to the health, besides being the nurseries of an innumerable host of frogs and mosquitoes....

Batavia, though not of an extraordinary size, nor embellished with buildings that are worthy of particular notice for elegance of design or magnificence of dimensions, may nevertheless be considered to rank among the neatest and the handsomest cities in the world. The ground plan is in the shape of a parallelogram, whose length from north to south is 4200 feet, and breadth 3000 feet. The streets are laid out in straight lines, and cross each other at right angles. Each street has its canal in the middle, cased with stone walls, which rise into a low parapet on the two margins. At the distance of six feet from this parapet wall is a row of evergreen trees, under the shade of which, on this intermediate space, are erected little open pavilions of wood, surrounded with seats, where the Dutch part of the inhabitants smoke their pipes and drink their beer in the cool of the evening. Beyond the trees is a gravelled road from thirty to sixty feet in width, terminated also on the opposite side by a second row of evergreens. This road is appropriated for the use of carriages, horses, cattle, and, as particularly pointed out by proclamation, for all *slaves*, who are strictly prohibited from walking on the flagged causeway in front of the houses, as they are also from wearing stockings and shoes, in order that their naked feet may be the means of making their condition notorious. This *trottoir* or footway is at least six feet wide; and as the breadth of the canals is generally the same as that of the carriage road, the whole width of the Batavian streets may be considered to run from 114 to 204 feet; and the city is said to contain twenty of such streets, with canals in the middle, over which they reckon about thirty stone bridges. The trees that embellish the streets are of different kinds, but the most common are two species of *Callophyllum*, called by botanists the *Inophyllum*, and the *Calaba*, the *Canarium Commune*, or canary-nut tree, the *Guettarda Speciosa*, with its odoriferous flowers, and the free, elegant, and spreading tamarind tree....

If a stranger should happen to make his first entrance into the city of Batavia about the middle of the day, he would be apt to conclude it deserted by the inhabitants. At this time the doors and windows are all shut, and not a creature, except perhaps a few slaves, is stirring in the streets. But if he should enter the city in the morning or the evening, his eye will not be less attracted by the vast crowds of people moving about in the principal streets, than by the very great variety of dress and complexion which these crowds exhibit. Here he will at once behold every tint of colour, except that of rosy health, from the pallid hue of the sickly European, through the endless shades of brown and yellow, to the jetty black of the Malabar; and the dresses of the several nations, both as to fashion and materials, are as various as their colour and cast of countenance. That class of men which bears a complete sway over the island is by much the least numerous; it is even rare to see a single *wel edele hoog gebooren Hollander, a right honourable high-born Dutchman*, condescending to walk the streets. 'Nothing from Europe,' he observes, 'but Englishmen and dogs walk in Batavia.' Whenever he has occasion to take this kind of exercise, he puts on his full dress suit of velvet, and is attended by a suitable retinue of slaves: sensible how very necessary it is, where power is but ideal, to put on an imposing appearance. But the Armenians, the Persians, and the Arabs, always grave and intent on business; the half-cast merchants from the different ports of Hindostan; and, above all, the Chinese, some in long sattin gowns and plaited tails reaching almost to their heels, and others crying their wares to sell, or seeking employment in their several professions, dressed in large umbrella hats, short jackets, and long wide trowsers; the Javanese loitering carelessly along, as if indifferent to every thing around them; the free Malays, with half-averted eye, looking with suspicion on all who come across them; and slaves, from every nation and country of the East, condemned to trudge in the same path with the carriages:—all these, in the early and latter parts of the day, may be seen bustling in crowds in the streets of Batavia....

[*Feasting with the Dutch.*]
We had scarcely set foot in the house when a procession of slaves made its appearance, with wine and gin, cordials, cakes and sweetmeats; a ceremony that was repeated to every new guest who arrived. After waiting a couple of hours

'... the dresses of the several nations, both as to fashion and materials, are as various as their colour and cast of countenance.' (From M. T. H. Perelaer, *Het Kamerlid van Berkenstein in Nederlandsch-Indie*, Leiden, 1888–9)

the signal for dinner was given by the entrance of three female slaves, one with a large silver bason, the second with a jar of the same metal filled with rose water for washing the hands, and the third with towels for wiping them. The company was very numerous and, the weather being remarkably close, the velvet coats and powdered wigs were now thrown aside, and their places supplied with short dimity jackets and muslin night-caps. I certainly do not remember ever to have seen an European table so completely loaded with what Van Weegerman was pleased to call *poison* and *pestilence*. Fish boiled and broiled, fowls in *curries* and *pillaws*, turkies and large capons, joints of beef boiled and roasted and stewed, soups, puddings, custards, and all kinds of pastry, were so crowded and jumbled together that there was scarcely any room for plates. Of the several kinds of dishes there was generally a pair: a turkey on one side had its brother turkey on the other, and capon stared at capon. A slave was placed behind the chair of each guest, besides those who handed round wine, gin, cordials, and Dutch or Danish beer, all of which are used profusely by the Dutch under an idea that, by promoting perspiration, they carry off in some degree the effects of the poison and pestilence. After dinner an elegant desert was served up of Chinese pastry, fruits in great variety, and sweetmeats. There were not any ladies in company. Van Weegerman being a bachelor had no females in his house, except his haram of slaves amounting to about fifty in number, assorted from the different nations of the East, and combining every tinge of complexion from the sickly faded hue of a dried tobacco leaf to the shining polish of black marble. A band of Malay musicians played in the viranda during dinner....

From table the Dutch part of the company retired to their beds, in order to recover, by a few hours sleep, the fatigues of eating and drinking, and to prepare for those of a far more serious meal which was to follow. The dinner, in fact, is considered only as a whetter of the appetite for supper....

[*The nine o'clock flower.*]

I have somewhere met with an observation, that an Englishman in building a house first plans out the kitchen; and a Dutchman the necessary. But the Dutch in Batavia, like the good people of Edinburgh, have contrived to dispense with conveniencies of this kind, for which I have heard two different reasons assigned: one is, that the heat of the climate would operate so as to create a putrid fever in the city; and the other, that the great bandicoot rat ... would infest the temple in such a manner as to render the resort to it unsafe, especially for the male sex: the first is absurd, the last ridiculous. Instead, however, of such places of retirement they substitute large jars, manufactured for the occasion in China, narrow at top, low, and bulging out in the middle to a great width. These jars remain undisturbed, in a certain corner of the house, for twenty-four hours; at the end of which time, that is to say at nine in the evening, the hour when all the parties usually break up and return to their respective homes, the Chinese *sampans* or dirt boats begin to traverse the canals of the city. At the well known cry of these industrious collectors of dirt, the slaves from the opposite houses dart out with their loaded jars, and empty their contents in bulk into the boats. In this manner the Chinese scavengers, paddling in their sampans along the several canals, collect from house to house, for the use of their countrymen who are the only gardeners, 'the golden store'. Such a custom, in such a climate, can be no less injurious to health than it is indecent and disgusting. But the Dutch appear to be as insensible of the one as they are reconciled to the other. If they happen to catch a passing breeze charged with the perfume of these jars, they coolly observe, '*Daar bloeit de foola nonas horas—the nine o'clock flower is just in blossom.*'

John Barrow, *A Voyage to Cochinchina in the Years 1792 and 1793* . . ., London: T. Cadell and W. Davies, 1806, pp. 171–2, 174–5, 202–3, 206–7, 213–14.

8
Thomas Stamford Raffles Discovers Java's Antiquities, 1815

When the English East India Company seized Java in 1811—to remove it from the clutches of Napoleon, who had placed his brother Louis on the throne of Holland in 1806—thirty-year-old Thomas Stamford Raffles was appointed head of the occupying British government. As Lieutenant-Governor, Raffles strove to reform the governance of the island according to his modern 'liberal' lights and, through scholarship, to bring Java comprehensively within the sphere of the known world. Tirelessly, he and his many assistants collected information on the island's flora and fauna, its languages, arts, religion, and history. Some of this knowledge he published in his encyclopaedic History of Java *in 1817, just one year after Britain returned the island to Holland following Napoleon's fall. In Java's Hindu and Buddhist temples and other 'antiquities', Raffles found evidence that the Javanese had once enjoyed a high civilization, which, in his view, had deteriorated under the impact of the later-coming religion of Islam—or Mahometanism, as he calls it. Raffles personally surveyed many of the ruins in 1815 and solicited lengthy reports about others from British military officers serving in Java as well as from the American naturalist Dr Thomas Horsfield. In* The History of Java, *Raffles reveals what he and his correspondents discovered. The following passages describe parts of the Loro Jonggrang, Borobudur, and Sari temple complexes of central Java, and the Singosari complex in east Java.*

THE natives are still devotedly attached to their ancient institutions, and though they have long ceased to respect the temples and idols of a former worship, they still retain a high respect for the laws, usages, and national observances which prevailed before the introduction of Mahometanism. And although some few individuals among them may aspire to a higher sanctity and closer conformity to Mahometanism than others, it may be fairly stated, that the Javans in general, while they believe in one supreme God, and that Mahomet was his Prophet, and observe some of the outward forms of the worship and observances, are little acquainted with the doctrines of that religion and are the least bigoted of its followers. . . .

Whatever of their more ancient faith may remain in the institutions, habits, and affections of the Javans, the island abounds in less perishable memorials of it. The antiquities of Java consist of ruins of edifices, and in particular of temples sacred to the former worship; images of deities found within them and scattered throughout the country, either sculptured in stone or cast in metal; inscriptions on stone and copper in ancient characters, and ancient coins.

The antiquities of Java have not, till lately, excited much notice; nor have they yet been sufficiently explored. The narrow policy of the Dutch denied to other nations facilities of research; and their own devotion to the pursuits of commerce was too exclusive to allow of their being much interested by the subject. The numerous and interesting remains of former art and grandeur, which exist in the ruins of temples and other edifices; the abundant treasures of sculpture and statuary with which some parts of the island are covered; and the evidences of a former state of religious belief and national improvement, which are presented in images, devices, and inscriptions, either lay entirely buried under rubbish, or were but partially examined. Nothing, therefore, of the ancient history of the people, of their institutions prior to the introduction of Mahometanism, of their magnificence and power before the distraction of internal war and the division of the country into petty contending sovereignties, or of their relations either to adjacent or distant tribes, in their origin, language, and religion, could be accurately known or fully relied on. The grandeur of their ancestors sounds like a fable in the mouth of the degenerate Javan; and it is only when it can be traced in monuments, which cannot be falsified, that we are led to give credit to their traditions concerning it. . . .

In addition to their claims on the consideration of the antiquarian, the ruins at two of these places, Brambanan [Prambanan] and Boro Bodo [Borobudur], are admirable as majestic works of art. The great extent of the masses of building covered in some parts with the luxuriant vegetation of the climate, the beauty and delicate execution of the separate portions, the symmetry and regularity of the whole, the great number and interesting character of the statues and bas-reliefs, with which they are ornamented, excite our wonder that they were not earlier examined, sketched, and described. . . .

Chandi Loro Jongrang; or Temples of Loro Jongran.

These lie directly in front (north) of the village of Brambanan, and about two hundred and fifty yards from the road, whence they are visible, in the form of large hillocks of fallen masses of stone, surmounted, and in some instances covered, with a profusion of trees and herbage of all descriptions. In the present dilapidated state of these venerable buildings, I found it very difficult to obtain a correct plan or description of their original disposition, extent, or even of their number and figure. Those that remain, with any degree of their primary form or elevation, are ten, disposed in three lines, running north and south. Of those on the western line, which are far the largest and most lofty, that in the centre towers high above the rest, and its jutting fragments lie tumbled about over a larger area. Nothing can exceed the air of desolation which this spot presents; and the feelings of every visitor are attuned, by the scene of surrounding devastation, to reflect, that while these noble monuments of the ancient splendour of religion and the arts are submitting, with sullen slowness, to the destructive hand of time and nature, the art which

Temple at Prambanan in Raffles's time. (From Thomas Stamford Raffles, *The History of Java*, London: Black, Parbury, and Allen, 1817)

raised them has perished before them, and the faith which they were to honour has now no other honour in the land....

The largest temple, apparently about ninety feet in height, is at present a mass of ruin, as well as the five others connected with it; but ascending to its northern face, over a vast heap of stones fallen from it and the third temple, at the height of about thirty feet you reach the entrance: the whole is of hewn stones, fitted and morticed into each other, without rubbish or cement of any kind. Directly in front of the doorway stands the image of Loro Jongran [Loro Jonggrang]....

The apartment in which this image and some other sculptured stones are placed, rises perfectly square and plain, to the height of ten feet, and there occurs a richly carved cornice of four fillets, a single stone to each. From this rises the roof in a square pyramid, perfectly plain or smooth, for ten feet more.

Proceeding over the ruins round to the west face of this building, you pass the intermediate angular projection, carved alternately in a running flower or foliage, ... and with small human figures of various form and attitude in compartments, above representations of square pyramidal temples, exactly like those on so many of the entablatures of Boro Bodo, and similar, I understand, to the Budh temples of Ava, &c. &c., the whole extremely rich and minute beyond description. The western doorway is equally plain with the former, and the entrance is still lower. The apartment is ten feet two inches square, apparently more filled up (that is, the floor raised higher than the other), but in all other respects exactly the same. In front is seated a complete Ganesa [four-armed elephant; Hindu god of prosperity], of smooth or polished stone, seated on a throne: the whole a single block, five and a half feet high and three wide. In his hands he has a plantain, a circlet of beads, a flower, and a cup to which the end of his proboscis is applied: a hooded snake encircles his body diagonally over the left shoulder. His cap is high, with a death's head and horned moon in front, and as well as his necklaces, waistband, armlets, bracelets, anklets and all his habiliments, is profusely decorated. The only damage he appears to have sustained is in losing all but the roots of his tusks.

The Javans to this day continue to pay their devoirs to him and to Loro Jongran, as they are constantly covered with turmerick, flowers, ochre, &c....

Remains of an Ancient Hall of Audience, &c. at Kali Bening. The temple which I have just described stands close to the north side of the village of Kali Bening, east of which is the river of that name; and as I had never before heard of any thing further in this quarter, I fancied my work was over. I was, however, most agreeably surprised, on being told by my Javan guides that there was something more to be seen directly south of the village behind us. We accordingly passed through it, and barely one hundred and fifty yards from the temple, in a high sugar-cane and palma christi [castor oil] plantation, we came suddenly on two pair[s] of very magnificent gigantic porters, all facing eastwards, each having stood about twelve feet from the others. The pedestals of all these statues are nearly covered, or rather entirely sunk into the ground. The height of each figure, from the top of the pedestal, is five feet one inch and a half, and breadth at the shoulders three feet six inches. They are generally much better executed, defined, and consequently more marked and striking in their appearance, than those I had seen. The countenance is much more marked and expressive, the nose more prominent and pointed, the eyebrows meeting in a formidable frown.... Behind the second pair of porters, or west of them, is a heap of ruins of brick and mortar, which proved on examination to be the remains of an ancient hall of audience or state, originally standing on fourteen pillars, with a verandah all round it standing on twenty-two pillars. The porters guarded this building exactly in the centre of its eastern front: the nearest pair scarcely thirty feet distant from it. The greatest length of the building was east and west. The inner apartment over all gave forty-seven feet in length, including the pillars: the width of the hall was twenty-eight feet and a half in the same way. A verandah, of twelve feet and a half wide all round over the pillars, surrounded the hall.

It struck me forcibly, that the house at Kali Sari was formerly the residence of some great Hindu Raja of Java; the superb temple at Kali Bening, the place of his devotions and prayers; this hall, a little south of it, that of state or audience, perhaps also of recreation after his devotions. Other ruins of brick-work, without any mixture of stone, were close by, and perhaps served as out-houses.

Boro Bodo.
In the district of Boro, in the province of Kedu, and near to the confluence of the rivers Elo and Praga, crowning a small hill, stands the temple of Boro Bodo, supposed by some to have been built in the sixth, and by others in the tenth century of the Javan era. It is a square stone building consisting of seven ranges of walls, each range decreasing as you ascend, till the building terminates in a kind of dome. It occupies the whole of the upper part of a conical hill, which appears to have been cut away so as to receive the walls and to accommodate itself to the figure of the whole structure. At the centre, resting on the very apex of the hill, is the dome before mentioned, of about fifty feet diameter; and in its present ruinous state, the upper part having fallen in, only

about twenty feet high. This is surrounded by a triple circle of towers, in number seventy-two, each occupied by an image looking outwards, and all connected by a stone casing of the hill, which externally has the appearance of a roof.

Descending from thence, you pass on each side of the building by steps through five handsome gateways, conducting to five successive terraces, which surround the hill on every side. The walls which support these terraces are covered with the richest sculpture on both sides, but more particularly on the side which forms an interior wall to the terrace below, and are raised so as to form a parapet on the other side. In the exterior of these parapets, at equal distances, are niches, each containing a naked figure sitting cross-legged, and considerably larger than life; the total number of which is not far short of four hundred. Above each niche is a little spire, another above each of the sides of the niche, and another upon the parapet between the sides of the neighbouring niches. The design is regular; the architectural and sculptural ornaments are profuse. The bas-reliefs represent a variety of scenes, apparently mythological, and executed with considerable taste and skill. The whole area occupied by this noble building is about six hundred and twenty feet either way....

The whole has the appearance of one solid building, and is about a hundred feet high, independently of the central spire of about twenty feet, which has fallen in. The interior consists almost entirely of the hill itself....

Ruins at Singa Sari, &c. in the District of Malang.

We first proceeded from Pasuruan to Lawang, mounting our horses at the ruins of a fort, which for some time withstood the Dutch arms on their first taking possession of these districts. Further on, between Lawang and Malang, the scene of a famous battle fought at that time was pointed out to us. The family of the present regent were first appointed to the office for services rendered on that occasion. The road from Pasuruan to Lawang lay principally through forests, in which we observed the waringen to predominate.

On the next morning we visited the ruins of Singa Sari, which are situated a few paces within the entrance of a teak forest, about four miles from Lawang, and to the right of the high road leading to Malang.

The first object which attracted our attention was the ruins of a chandi or temple. It is a square building, having the entrance on the western side: its present height may be about thirty feet. Over the entrance is an enormous gorgon head, and a similar ornament appears originally to have been placed on each of the other sides of the building, over the niches, which correspond with the entrance on the western side. In one of these niches we observed an image lying flat on the ground, with its head off; in another, the pedestal of an image, which we were informed had been taken away by Mr Engelhard;[1] and where the traces of a third niche appeared, the stones had been removed, and a deep hole dug, so as to disfigure, and in a great measure demolish, this part of the building. This was also attributed to Mr Engelhard's agents.

On entering the chandi, to which we ascended by stones which had evidently been once placed as steps, we observed a very deep excavation, and a large square stone upset and thrown on one side. We ordered it to be filled up and the large stone replaced. There was a round hole passing completely through the centre of this stone, which, whether it had been an altar, the pedestal to some image, or a yoni [stylized vagina], we could not ascertain.

Without the building, on part of the ruins which appeared to have been the lower terrace, we noticed two porters, with clubs in their hands, resting on the shoulder. The features were entirely defaced, and the images rude; but we easily recognized their similarity to the porters at Brambanan. They were, however, not above three feet high.

The devices, ornaments, and general style of this temple are not very different from those of the great temple at Brambanan: the cornices and mouldings are no less rich and well executed. The external form of the building may differ, but the recess, or chamber within, seems on the same principle. There is no inlet for the light from above.

Proceeding a short distance further into the forest, we found several images of the Hindu mythology, in excellent preserva-

[1]Mr Engelhard, Raffles tells us later, was the former governor of Semarang in central Java: 'In the garden of the residency of that state, several very beautiful subjects in stone were arranged, brought in from different parts of the country,' *The History of Java*, p. 55.

tion, and more highly executed than any we had previously seen in the island. In the centre, without protection from the weather, was the bull Nandi, quite perfect, with the exception of the horns, one of which was lying by the side of it. This image is above five feet and a half long, in high preservation, and of excellent proportion and workmanship.

Near the bull, and placed against a tree, is a magnificent Brahma. The four heads are perfect, except that there is a mutilation about the nose. The figure is highly ornamented, and more richly dressed than is usual....

At the distance of about a hundred yards from this spot, we were conducted to a magnificent Ganesa of a colossal size, most beautifully executed, and in high preservation. The pedestal is surrounded by skulls, and skulls seem used not only as ear-rings, but as the decoration of every part to which they can be applied. The head and trunk are very correct imitations of nature. The figure appears to have stood on a platform of stone; and from the number of stones scattered, it is not improbable it may have been inclosed in a niche or temple.

Still further in the wood, at a short distance, we found another colossal statue, of the same stamp as the porters at Brambanan. This statue was lying on its face at the entrance of an elevated stone terrace: but the people having excavated and cleared the earth around, we were enabled distinctly to examine the face and front. It measures in length about twelve feet, breadth between the shoulders nine feet and a half, and at the base nine feet by five, and is cut from one solid stone. The figure is represented as sitting on its hams, with the hand resting on each knee, but no club, although it is not impossible it may have been broken off. The countenance is well expressed and the nose prominent; but this feature, as well as the mouth and chin, have suffered injury from partial mutilation.

The statue seems evidently to have fallen from the adjacent elevated terrace, which is about eighteen feet high in its present dilapidated state, and is built of stones, the upper ones being immense slabs of five feet by four, and three feet thick. A second figure of the same dimensions was afterwards found in the vicinity; these were no doubt porters who guarded the entrance to these temples.

Having visited all that could be traced in the vicinity of Singa Sari, we proceeded on to Malang....

Thomas Stamford Raffles, *The History of Java*, London: Black, Parbury, and Allen, 1817, Vol. 2, pp. 2, 5–6, 11–14, 28–30, 41–3.

9
Charles Walter Kinloch Takes a Cure, 1852

'Never met with a clearer case in my life,' says Charles Walter Kinloch's doctor, diagnosing dyspepsia. 'You must get to sea without delay.'[1] *Kinloch, a resident of British Bengal, heads for Java. His itinerary there—taking in Batavia, Buitenzorg (Bogor), Bandung, Borobudur, and other sights along the way—prefigures those of many a traveller to come. Kinloch is truly a tourist, albeit a grumpy one. Crossing the island by carriage, he witnesses tea plantations, coffee gardens, and other signs of colonial enterprise such as the good roads and efficient horse-posting services. The Javanese (who, he writes, have 'coarse features' and are 'awkward and ungraceful') figure little in his observations.*[2] *And when he laments, 'How much a person loses by being unacquainted with the language of the country through which he is travelling,' he means Dutch!*[3]

THE approach to Batavia is cheerless in the extreme. The town is situated amid a low marshy jungle, the very hot bed of malaria; and as the ships lie out at a considerable distance from the shore, in order to escape the ill effects of the baneful land-wind, landing is at all times a tedious affair. The town is reached by a canal, which flows through it, and

[1][C. W. Kinloch], *De Zieke Reiziger; or Rambles in Java and the Straits in 1852. By a Bengal Civilian*, London: Simpkin, Marshall and Co., 1853, pp. 1–2.
[2]Ibid., p. 73.
[3]Ibid., p. 58–9.

for several miles into the interior of the country. The climate of Batavia, as is well known, has always proved most deadly to the European constitution; and even at the present day, it is only the native portion of the community that can remain in the town with impunity after night fall. The European population reside entirely in the country, at a distance of three or four miles from the town; and the merchants and others who have business to transact, go up to their offices daily at an early hour of the morning, and by three o'clock in the afternoon all business has ceased, and every office in Batavia is closed.

We did not reach our hotel till nearly nine o'clock at night, having had the usual amount of trouble at the Custom House that falls to the lot of those who are strangers in the land, and are unacquainted with the language of the people.

It is difficult to say which is the best hotel at a place where all are bad. We were advised to reside at The Rotterdamsche; but we should have done better, as we subsequently discovered, had we selected the Hotel der Nederlanden. Considering the very large number of English residents at Batavia, and the constant influx of English visitors from different quarters of the globe, it is strange there should be no hotel of a purely English character; but the whole are essentially Dutch; and the English visitor, therefore, so long as he may reside in Java, must learn to live like a Dutchman, or he will chance to die of starvation. He must take his breakfast at 6 a.m., or not at all. He must prepare to dine at noon. If it be his principle 'when at Rome to do as the Romans do', he will then go to bed for an hour or two, take a cup of tea on rising, and dress for the day about five o'clock in the afternoon; he will then be ready for a second dinner at eight o'clock; and from that hour until midnight, he may amuse himself at the billiard table, or if he prefer society, he may pay his 'devoirs' to some of the many fair ladies of Batavia....

Dutch cooking would never suit an English stomach; it is not only not wholesome, but it is even worse, it is disgusting, the predominating features of it being acids and rancid butter. In every Dutch dish there is a disagreeable excess of these adjuncts. Another curious feature connected with it is, that, with the exception of the soup, which is served up upon the boil, the rest of the dinner is allowed to become quite cold before it is eaten; the amount of caloric that is necessarily diffused over the coats of the stomach, through the introduction of the former, renders it expedient, we were told, that the temperature of the other dishes should be proportionably lowered; accordingly, the whole of the viands are invariably set out on the table about half an hour before the dinner is announced, in open flat dishes, in order that they may have time to cool before they are handed round. There are no separate courses at a Dutch dinner, the whole meal, including the desert, being displayed at once upon the table, so that, on taking his seat, a person may literally be said to see his dinner.

In their mode of eating, the Dutch have never studied refinement; and even at the table of the highest in the land, it is customary for persons of either sex to employ their knives on offices which with us are usually performed with the spoon or fork. The use of a butterknife or a saltspoon is unknown in Java; and even at the Governor General's table, these useful appurtenances of the dinner table are not deemed necessary.

The Batavians are very fond of gaiety, and strive to forget the depressing effects of the climate in the indulgence of one continued round of balls and dinner parties. The opera (French) is a very passable one; we were there on a benefit night, and heard Halevy's 'La Juive', a difficult selection, but to which the company, we thought, did very fair justice....

The Batavian ladies dress with taste, when they do dress; but this necessary operation is seldom performed until a very late hour of the day. The morning dress of a Batavian lady consists merely of a pair of wide silk trousers, or a coloured Malay petticoat, and that indispensable upper garment termed a shift; the naked feet are then carelessly thrust into a pair of Chinese slippers, and in this guise, with their hair uncombed and floating down their backs, scores of ladies may be seen every day of the week driving along the public road, or walking before their houses in familiar conversation with their friends of either sex. We never once remarked anything like confusion or awkwardness on the part of any lady on being encountered so dressed, or rather so undressed.

On the first morning after our arrival, we came suddenly upon a lady thus habited. She was proceeding across the quadrangle of the hotel to the bath room, followed by a female servant, bearing towels and certain other 'et ceteras' that are used by ladies when performing their ablutions. There was no escape for ourselves or the lady; but though somewhat disconcerted ourselves, we were surprised to find that the lady looked wholly unconcerned. We were afterwards told that had we studied Batavian etiquette, we should have taken off our hats to the lady on passing her. But it is not only by an indecent display of their persons that the Dutch ladies evince their disregard for delicacy and propriety: they are also complete strangers, in our humble opinion, to that innate good breeding, and that natural good taste, which are so characteristic of the sex in England....

But if the ladies are deficient in delicacy, the gentlemen are complete strangers to the feeling. It is a common sight at the leading hotels of the capital, to see a gentleman sitting, in the middle of the day, in his night dress in the public verandah of the hotel, undergoing the process of shaving, or performing certain ablutions which more properly belong to the bath room. As we now write, there are sitting in front of our windows, and not thirty yards from the door of our apartment, two Dutch gentlemen, the one in sky blue drawers and a shirt, the other in slate coloured drawers, but with no shirt; the former is solacing himself with a cheroot, whilst his companion is apparently engaged in qualifying for the office of a chiropodist.

If there is one class of persons more than another in Java, against whom we have cause to entertain feelings of hostility, that class is the washermen. We arrived on the island with a perfectly new wardrobe: we left it with scarcely a sound garment. We could pardon, perhaps, the natural indolence which sometimes induces this class of servants to refuse to wash for you on any terms; but we cannot so readily overlook that peculiar feature in the character of the Javanese Dhobi [washerman], which leads him to beat all your garments into shreds for the mere love of the thing. One of their most objectionable practices is to starch your shirts to such a degree, that, even if you have the good fortune to find your way into one of them, every movement you may make is attended with the greatest personal discomfort. On remonstrating with our washerman against this abominable custom, we were told that the gentlemen all over Java wore their shirts stiffly starched, as this garment was seldom worn inside of the trowsers, except of an evening. As this statement was corroborated by what we had already observed ourselves, we could say nothing further on the subject, and yielded the point accordingly....

Travelling in Java is very expensive, the average cost per mile being two Java rupees, or one rupee eight anas of our Calcutta currency. The roads are excellent, as are the horses also, the actual travelling pace of the latter being upwards of ten miles an hour. Post horses are only obtainable by application to the Government, whose sanction must be first procured ere the visitor will be permitted to quit the capital. The usual step on arriving at Batavia, is for the stranger to submit a petition to the Governor General, praying for permission to visit the interior. It is a troublesome form, but it is one that must be observed. Sanction is given, as a matter of course, unless, indeed, some special cause should exist for its being withheld. The applicant then receives a passport, which holds good for a twelvemonth, and for which he has to pay the cost of the stamp, only two and a half rupees....

On the morning of the 9th of June, we found ourselves seated behind four neat little posters, who whisked us out of the yard of the 'Rotterdamsche' at a pace that promised to bring us to Buitenzorg within the four hours which are usually allowed for reaching that place. The Java ponies seemed to think nothing of our heavily laden carriage, as they galloped up and down hill at the uniform rate of ten miles an hour.

There was something so exhilirating in the rapid pace, and the mountain air seemed so pure and fresh, that we already felt better than we had done for months....

We were joined at the hotel, two days after our arrival at Buitenzorg, by two Indian friends, who, like ourselves, had come to Java in quest of health. We had strolled out one evening with these gentlemen for a short walk, as we thought, in the environs of the village, when we were unexpectedly overtaken by one of those sudden thunder storms which are peculiar to the climate of this place. We had for some time been walking, as we imagined, in the direction of the hotel, and though the charms of the walk had been effectually destroyed by the storm that was now raging around us, still we were all comforted by the feeling that in a very few minutes we should be under shelter. Thus assured, we continued to walk on through the torrents of rain that were now descending upon us. At length, as night began to fall, and as neither on our right hand nor on our left, nor immediately in front of us, could we distinguish anything that seemed to indicate the proximity of a populous village, it occurred to us for the first time that we were not upon the road to Buitenzorg. None of the party could speak a word of Malay, and it was not likely that any of the peasants whom we might meet upon the road could speak any other language. We, therefore, merely repeated the word 'Buitenzorg', as we earnestly inquired of each passer by whether we were on the road to that village. 'Ya, ya,' was the invariable answer we got, reiterated, too, in so decided a tone, as to leave little doubt on our minds that we were pursuing the right direction.

On we went, therefore, in spite of the mile stones, which plainly indicated that each onward step we took was only leading us further away from the place of which we were in quest. Foolishly, however, distrusting those silent counsellors, and relying with a blind confidence on the 'ya ya' of the Malay peasant, we still walked on. At length we secured a guide from a hovel at the roadside, and endeavoured by signs and other expedients to make him understand that we wished to get to Buitenzorg. Under this man's guidance, we continued to walk for about two miles further, in the heaviest rain to which we have almost ever been exposed. When, at length, we had the happiness to distinguish lights, we shortly afterwards entered a village, and were taken to the house of a Dutch gentleman, who informed us that we were at Grimagah, a village distant some seven miles from Buitenzorg. He likewise told us that the Malays did not know the latter place under the name of Buitenzorg, but that they called it Bogor....

June 18.—'All ready, Sir!' screamed our Malay courier; crack, crack, go the several whips of coachman and Syces [grooms], and away spring the ponies at full gallop, up and down hill, it's all the same to them; the pace is too rapid to admit of their feeling the ascents, no matter how steep. Why, by all that's marvellous, here we are at the post station, and only twenty seven minutes doing the six miles from Buitenzorg! If this be the usual rate of travelling on the island, there is little grass can grow beneath the hoof of a Java pony. What a glorious morning we have, too; how fresh and pure the air, and how delicious the fragrance of the countless wild flowers that adorn the road side banks, scenting the air with a perfume which reminds one of the syringa of our English gardens. Then what a glorious scene lies before us; immediately in front is the richly wooded Megamendon [Megamendung], over which, some four hours hence, our carriage will pass, at an elevation of four thousand three hundred feet above the sea.

A little to the right, and already enveloped in mist, rises the lofty Simoet. On the extreme left, and occasionally displaying its crest through the white fleecy clouds that are sporting upon its summit, stands the noble Salok [Salak], at an elevation of seven thousand feet. Beneath us, and now gradually receding from view, are the many picturesque, though low hills, which almost encircle the town of Buitenzorg; and far, far away, and for many a mile, stretch the verdant plains which lie betwixt the districts of Buitenzorg and Batavia.

The stages are not long, averaging from five to six miles; but they are quite long enough, considering the whole distance is done at the gallop. At each post, an open shed with verandahs on either side of it, is built across the road. Under this shed the operation of changing horses is effected; and whilst this is going forward, the traveller will probably have not failed to observe that the considerate solicitude of the Government for the passengers' comfort does not end here, but that other buildings have also been constructed by it, even more conducive to the travellers' comfort than the one already noticed.

At Tjiserooa [Cisarua] the road becomes too steep and too stony for horses only, and from this point, and until the summit of the Megamendon is attained, it is usual to employ buffaloes to aid the horses in the ascent—two, three, or four pairs at a time, according to circumstances.

The view of the Prianger districts from the summit of the pass is magnificent. We had had a smart shower of rain before reaching the top, but it cleared just in time to give us a splendid prospect of the plains below us. A descent of about a thousand feet bring[s] the traveller to Tjipanas [Cipanas]; here there is a private bungalow belonging to the Governor General, a small botanical garden, and some hot springs. A further descent of two thousand feet, and the traveller reaches Tjanjore [Cianjur], the head quarters of the Resident or chief civil authority of the Priangen.

The village is prettily situated, and it has an air of cleanliness and comfort about it, which we may look for in vain in any of the villages of continental India. The hotel is not well situated, and has a very gloomy look; the bed rooms, too, are dirty and ill ventilated; the dinner was quite in character, and was uneatable; the only object that one could look upon with anything like complacency, was the good humoured face of the Dutch landlady. She could speak a little English, she told us, but this little was scarcely enough to admit of a conversation being long maintained, being limited to this: 'very'; 'how d'ye do'; 'yes'; 'no'; and 'good morning'.

We were not sorry to emerge from the gates of this gloomy caravansery at an early hour on the following morning in progress to Bandong [Bandung], where it was our intention to sojourn some considerable time, in order to enjoy its far famed climate, and to see the lions of that highly favoured neighbourhood. The distance from Tjanjore is forty two English miles, and is usually performed in six hours. The road runs through a most picturesque part of the country, and in several places it is necessary to employ buffaloes, as on the previous day, to drag the carriage up the steeper hills.

About fifteen miles from Tjanjore, an exceedingly abrupt descent brings the traveller to a tributary of the Tjeetaram [Citarum]; the inclination of the road is here so great, that it is necessary to attach a treck tow, or leathern rope, to the hind part of the carriage, upon which a strong pull is maintained by some twenty or thirty Coolies, in order to prevent too rapid a descent of the carriage down the hill. The approach to this stream is very beautiful, but the view in the descent to the river itself is still more beautiful. The river is crossed by a punt without trouble or delay; and immediately on gaining the opposite side, four powerful buffaloes are yoked to the carriage, which in the course of a few minutes is safely transported to the top of the opposite bank.

'The river is crossed by a punt without trouble or delay. . . .' (From M. T. H. Perelaer, *Het Kamerlid van Berkenstein in Nederlandsch-Indie*, Leiden, 1888–9)

We consider this ascent from the Tjeetaram to be the only really dangerous part of the road between Batavia and Bandong; and the most courageous person might, perhaps, be pardoned for feeling somewhat nervous whilst engaged in travelling up it, feeling as he must do, that his safety, and even his life, are dependant for a few minutes upon the strength of a piece of untanned buffalo's hide. The last sixteen miles into Bandong are over a comparatively level road; buffaloes are no longer needed, and away spring the Java ponies at their customary pace, bringing the traveller to the door of the hotel in a few minutes over the hour. . . .

The environs of the village [Bandung] are almost exclusively occupied by coffee gardens, each plantation being fenced in with a closely cut hedge of the scarlet hibiscus, which here grows in the greatest luxuriance. It is not until one has fairly left the village and ascended one or other of the heights above Bandong, that a good view of the village is obtained. At an elevation of a few hundred feet above the town, the traveller will be rewarded, if the weather be clear, with a fine panoramic view of the surrounding hills, and of the rich valley below him, in the centre of which he will see the village embosomed in its numerous coffee gardens, and luxuriant with a perpetual verdure.

Made the acquaintance of Mr L—, a gentleman resident in this neighbourhood. This gentleman was formerly in the service of Government, and held the situation of Resident in one of the eastern districts of the island; this appointment, however, he resigned some years ago, preferring the independent position of a private country gentleman to the highest official post under Government.

Mr L owns one of the most lucrative tea plantations in Java, and the property, we understand, is increasing yearly in value. He holds a certain number of acres from Government, under a contract to supply yearly to the State at fixed prices, as much tea as his land may be found capable of producing. This is only the third year of the contract, and Mr L— has already under cultivation three hundred and fifty baos[4] of land, the gross yield of which may be estimated at two hundred thousand pounds. The average price per pound that the contractor receives for the several descriptions of teas, is seventy-five cents, or about one shilling and a half penny of our English currency, whilst the actual expense to him scarcely exceeds a half of this sum; the net profit, therefore, to the contractor, in this the third year of his contract, may be estimated at about £6250 sterling. The number of labourers, chiefly Javanese, at present employed upon the property, is upwards of one thousand seven hundred; but this number will, of course, increase as the cultivation is extended....

Bandong well deserves, we think, the character it bears of being the Montpelier of Java. During the fortnight we were there, the weather was truly delightful, the thermometer

[4]The bao is somewhat larger than the English acre.

never rising above 75° at the hottest period of the day, and frequently falling as low as 68° before sunrise. The town is situated at an elevation of two thousand two hundred and forty feet above the level of the sea, and there are several lofty mountains in its immediate vicinity; amongst the number, the Goonangago [probably Gunung Tangkuban Prahu] and the Goonangrang-rang [Gunung Burangrang], which rise respectively to the height of seven thousand five hundred, and six thousand eight hundred feet.

During the present and two succeeding months, which are considered the driest of the year, rain falls every second or third day, but it seldom lasts beyond an hour or two, and the soil quickly dries, so that the roads are never dirty, and are always free from dust. In such a climate as Bandong, the Indian invalid could not fail to gain health and strength, were it possible for him to meet with wholesome food; but the diet and the cookery are quite unsuited to an English taste; and to an invalid, they are perfect poison.

Fine air, picturesque scenery, and healthful exercise, may do something; but they will not do much, if bread, meat, butter, and fish, are left out of the scale; and none of these common necessaries of life are procurable within eighty miles of Bandong. . . .

July 1.—The drive from Bandong to Somadang [Sumedang], twenty nine miles, is exceedingly pretty, particularly that portion of the road that leads directly down to the Ising Koep Port. The view from the top of the hill, before you descend to this station, is highly picturesque, and would afford a pleasing subject for the pencil. There is a quiet rural beauty about Somadang that is to our mind peculiarly delightful. . . .

The hotel we found clean and comfortable, and our only regret was that we had not made arrangements for spending several days in this delightful village. . . .

July 2.—The scenery on the road from hence to Cheribon, fifty nine miles, is of a very varied character. During the first half of the journey, and until the river is crossed, there are several steep hills to be surmounted, from the summits of which the traveller may obtain some fine views of the country. After passing the river, the road runs along the lowlands, passing through extensive sugar farms, upon which several hundred Chinamen may be seen pursuing their various occupations with untiring industry, affording a marked contrast to the indolent Javanese, who take no thought beyond the present moment. Cheribon is a dismal looking place on the sea coast, with a miserable hotel, out of which we were glad to make our exit on the following morning, 'en route' to Tagal....

Tagal, July 3.—Neither the hotels nor the horses improve as we advance eastward. As regards the former, there is on the part of the owners a perceptibly increasing disregard for cleanliness, and a corresponding predilection for greasy cookery; and as regards the latter, there is on the part of the ponies, more particularly those of the Cheribon district, a marked aversion to leave home, which renders travelling upon this most uninteresting portion of the road far less agreeable than in most other parts of the island. The distance from Cheribon to Tagal is frequently performed in four hours; but our horses were not in good humour, and we were upwards of six hours on the road. Here we halt one day to enable the Resident to make arrangements for our progress by a cross mountain road to Banjoemas [Banyumas], and for our reception and entertainment at the Passengrang [*passanggrahan*] of Bomiajoe [Bumiayu].

Tents are not used in Java as with us in British India; but as some kind of accommodation is necessary for the officers of Government, when engaged in making the circuit of their districts, buildings of bamboo have been constructed at certain localities and distances in each district, at the public expense. These buildings are usually called a Passengrang, and are exceedingly comfortable....

An uninteresting drive of thirty eight miles brought us to Kubooman [Kebumen], a small district in charge of an Assistant Resident. We were very kindly received by the officer in charge, Mr Petel, who accompanied us in the evening to see a review of some Javanese troops, belonging to the native chiefs of the neighbourhood, who were practising a variety of evolutions for a public entertainment to be given by the Assistant Resident at the close of the Ramzan [fasting month]. Shortly after our arrival on the Parade ground, the Regent of Kubooman made his appearance, mounted upon a bright bay horse, small, but of exceeding strength, the trappings of which were all of wrought silver. No sooner was the figure of this important personage distinguished by the crowd of retainers and others assembled on the plain, than the whole living mass sank simultaneously to the ground, in token of the respect that was due to his superior rank. The Regent then rode slowly forward to a raised platform, that had apparently been erected for the occasion, and having dismounted from his horse, took his seat amongst the party of the Assistant Resident. No sooner had he seated himself, than a hundred human beings were seen to emerge from the prostrate crowd, and with their hams still resting on their heels, to shuffle themselves along the ground with surprising quickness, till having arrived in front of the platform, they ranged themselves in a semicircle before the Regent, still taking care not to quit the unbecoming and degrading posture above described.

No stranger can have been a week in Java without having occasion to notice the servile deference that is paid by the Javanese to superior rank. The Chinese evince their respect for rank by removing their hats whenever a superior is passing by; but the Javanese show their respect to him by suddenly sinking to the ground with their hams resting on their heels. The posture is as ungraceful as it is degrading. All orders are asked and received in this humiliating position, and no servant or other inferior will durst assume any other posture, whilst he is in the presence or within sight of a superior, even though that superior may not be his master. The custom is so intimately mixed up with the institutions of the country, that it would be a difficult matter perhaps to effect its abolition; but we learn from Raffles that during the brief administration of the English, the practice was in some measure discontinued. . . .

The village of Salatiga stands at an elevation of eighteen hundred feet above the sea; and it furnishes a most agreeable retreat to the merchants and others resident at Samarang, from the almost unsupportable heat of that place.

The hotel at Salatiga is a mercantile building, being enclosed on every side by trees and bushes, which effectually exclude all circulation of the air. The sitting rooms reek with the fumes of tobacco, and have a dirty squalid look, which extends to everything about the establishment. The cookery was quite in character, and could only be fitly designated by the term disgusting.

Samarang, July 14.—Rancid butter, musty bread, unmanageably tough fowls, and the eternal fricandell, have now pursued us, step by step, over a distance of nearly five hundred miles; and though we can appreciate most fully the great natural beauties of the country, and can bear testimony to its delicious climate, and though we have experienced, but with one exception, unvaried kindness and attention at the hands of the European Residents, from the Governor General down to the humble hotel keeper, yet our unimproved health warns us that we must flee this beautiful island, from the utter impossibility of combining along with its salubrious climate a diet that is in any way suited to an invalid....

'How do you find the country?' is generally the leading question which every Dutchman puts to you on your first introduction to him. He will then, without giving you time to reply, proceed to answer his own query, by telling you that the country is the finest in the world, and the climate unrivalled. Without going quite so far as this, we are free to admit that we have never seen any country more highly favoured by nature than the Island of Java. Under the proverbially inert administration of the Dutch, however, but little progress has as yet been made in developing the vast resources of the country....

Had this island remained in British possession, it is probable that matters would have been very different from what they now are. The silly, vexatious passport system would have ceased to exist, travelling would have been made available to all classes, English capital and English enterprise would have destroyed all monopolies, and private competition would long since have lowered the expense of posting to such a rate as to place travelling within the reach of almost every class. As matters stand at present, under the Dutch Government, the rates for posting are so extravagantly high, being on an average two Java rupees per mile, as to put it out of the power of all save those who are in independent circumstances to see anything whatever of the country....

The Government is absurdly jealous of strangers, and though it rarely goes the length of refusing them permission to travel over the island, provided all the necessary forms have been observed by the applicant, yet sanction to visit the interior is never willingly accorded, and the authorities are always ready to avail themselves of any good ground for withholding the requisite permission. As it is, the high rate for posting charged by Government, as we have already remarked, amounts almost to a prohibition upon travelling; then again the many minute and troublesome forms that have to be observed by the intending traveller, added to the annoyance of the absurdly rigorous passport system, tend in no small degree to discourage travelling, even among those whose circumstances may admit of their indulging a wish of the kind.

We have not yet resided two months on the island, yet on no less than four separate occasions, have our trunks and other baggage been subjected to a rigorous examination by the Custom House authorities. Of the first examination, which was made on our first arrival in the country, we have no right to complain; but we do complain of, and protest against, the other three examinations, as quite unnecessary and as extremely vexatious. We had never once left the island, as our passport could testify; yet, because we found it convenient to return to Batavia by sea instead of by land, we were subjected to the annoyance of having our luggage searched at the port of embarkation, as also again on reaching the capital; and on leaving the island a few days afterwards for Singapore, we had to submit to the same annoyance for the fourth time in two months, in order that the Dutch Government might be certified that we had not carried away from the country more than the authorized amount of bullion....

As regards the suitableness of Java as a place of residence for the Indian invalid, who may merely need a temporary change of climate, or whose term of leave might not allow of his visiting the Cape or Australia, we are of opinion that within the wide range of what are called 'Indian limits', there is no climate to be found superior to that of Java, or one more easy of access to the invalid. The great drawback, however, to Java, as already remarked, consists in the want of proper accommodation and suitable food in those particular parts of the island where the climate is most salubrious and the scenery most attractive. Travelling, too, is not prosecuted with the same ease, and at the same trifling cost, with which it may be indulged in, in British India; but still we do not see why those whose means are not greatly circumscribed, should not during the dry season—viz. from June until October, march over this island in the same independent and delightful mode in which they have been accustomed to travel in our own provinces.

Provided with an accreditory letter to the Dutch Government, and without which the traveller could do nothing, nor move an inch in the country, the invalid visitor will be enabled to obtain any number of Coolies he may need for the transport of his baggage.

Let him bring along with him a couple of hill tents, and an up country tent pitcher; let him furthermore provide himself with a serviceable Mussulman cook, and we see no reason why he should not travel from one end of Java to the other with the same ease as if he were travelling in Bengal.

[C. W. Kinloch] *De Zieke Reiziger; or Rambles in Java and the Straits in 1852. By a Bengal Civilian,* London: Simpkin, Marshall and Co., 1853, pp. 30–42, 46–8, 50–4, 56–8, 64–5, 70–2, 74, 81–3, 96–9, 101–4, 106–8, 114–15.

10
J. W. B. Money Finds Java a Happy Spot, Indeed, 1858

Like Charles Walter Kinloch, J. W. B. Money was a Briton based in British India, and like Kinloch, he made a trip to Java to effect a cure—in this case for his wife. (Mr and Mrs Money were accompanied on their journey by their Portuguese maid, L'illustrissima Dona Carolina Maria de Pinto de Cruz.) Unlike Kinloch, however, barrister Money was uniformly impressed with Java which, he writes, acted 'like magic on an invalid'.[1] *Even notorious Batavia was 'one of the cleanest and prettiest of cities', all the nicer for lacking pompous, luxury-loving subjects such as the 'Baboos' of Calcutta, his hometown.*[2] *Indeed, in every respect Java*

[1] J. W. B. Money, *Java, or How to Manage a Colony,* London: Hurst and Blackett, 1861, Vol. 1, p. 19.

[2] Ibid., p. 5.

compared favourably to India, still reeling at the time of his trip from the effects of the disastrous Indian Mutiny of 1857. Money's positive view of Java was surely influenced by the rich company he kept, as depicted below in his accounts of horse-races and stag hunts with Java's high Dutch and Javan officials. Money's inaccurate treatise on the Dutch colonial administration, which takes up the vast majority of his long book about Java, influenced several later writers and helped to establish a reputation for the Dutch as model colonizers.

FOUR years' residence in Calcutta, and a habit of inquiring into Native ideas on different subjects, had made me tolerably acquainted with the general outline of things in India, and with the wishes and grievances of the Natives of Bengal, while the general discussion of Anglo-Indian theories at that time put me *au fait* as to the proposed alterations. In the middle of these discussions, my wife's health requiring change, we selected Java for a trip in the summer of 1858, more from hearing that it was a beautiful island, with a fine climate, easy travelling, and an opera, than with any idea of acquiring useful information from an examination of the Dutch colonial system.

Before leaving Calcutta I made inquiries, and found that the English information, on the subject of the Dutch East Indies, was generally limited to a recollection of Sir Stamford Raffles's account of Java in 1815.... I could find but one recent work, with the affected title of 'De Zieke Reiziger' (Dutch for 'The Sick Traveller'), or 'Rambles in Java in 1852 by a Bengal Civilian', containing descriptions of the goodness of the roads, the beauty of the scenery, the wonderful costume and alleged want of delicacy of the Dutch, both ladies and gentlemen, and the abominable Dutch cookery, which, according to this Bengal civilian's own account, had driven him from the island, leaving a great part of its beauties unexplored. This book is a source of constant annoyance to the English traveller in Java. The author, judging everything by Anglo-Indian ideas, writes of the manners and costumes of the Europeans there, who have adopted habits suited to a hot country, in a manner which has not only caused general indignation throughout the island, but has shut the doors of every house against English strangers, during the familiar hours of homely deshabille.

On our arrival at Singapore I renewed my inquiries, but our countrymen there seemed hardly better acquainted with the state of Java. They told me that the Dutch Colonial Government was a secret, monopolizing, and tyrannical Government, of which little was known, except that it was said to be hated by its Native subjects, who only wanted encouragement to throw off the Dutch yoke and to return to English rule.

We arrived in Java, therefore, expecting to find an oppressed and poverty-stricken people, with general marks of misgovernment by the Europeans, and of discontent among the Natives.

Batavia.—On landing at Batavia, the capital of Java, we found a state of things not easily reconcileable with these anticipations. A large and flourishing mercantile community of English, French, and Germans, with general comfort and apparent cheerfulness among the Natives in the town, somewhat contradicted the accounts we had heard. The lower classes, composed of Chinese, Javanese, and Malays, were evidently well off, but there are no Natives in Batavia living in the luxury, and displaying the pomp, of the rich Baboos in Calcutta.

Batavia itself is one of the cleanest and prettiest of cities. The lower or business part of the town, where the tide rises and falls in the canals, and which is near the seashore marshes, was formerly very unhealthy, and is still objectionable at night. The upper or European part, where the hotels, the clubs, the opera, and the concert rooms are situated, is some two miles from the Lower Town; the canals there are clear flowing streams; and epidemical disease is now rare, even in the Lower Town, and almost unknown in the Upper.

The frightful periodical mortality, which made Batavia a by-word under our rule, called for large sanitary measures. Such have been strictly carried out by the Dutch since their return, till continual cleaning, sweeping, and draining have made the most deadly city in the world an agreeable and healthy residence. . . .

After a short stay in Batavia, we proceeded to make a trip through the hilly interior of the western end of the island, among high mountains, still smoking volcanoes, grand and rugged scenery, variegated by the dense foliage and the beautiful flowers of the Java forests. The exhilarating, bracing air peculiar to the highlands of tropical countries, the diversified impressions produced by change of scene, and, more than all, the ease and comfort of travelling in Java, are admirably adapted to Indian invalids.

Java Posting.—The means of locomotion in Java are as good as they were on the Continent before railways were introduced, and the contrast with India in that respect cannot but strike an Englishman with surprise and regret. The Dutch Government takes a comprehensive view of the duties of the State in this respect, although, like that of India, it derives no direct benefit from the traffic which it thus encourages. In Java, all who are able to afford the expense post in their own carriages with government horses. The traveller pays for the horses at the post-office, the expenses, which are considerable, being about four shillings a mile. The chief local European official, the resident, regulates the traffic along the road, so as to secure to each set of post-horses six hours' rest between each journey. Notice of each traveller's movements is sent on beforehand to the other residents, that they may make such preparations as will prevent the traveller from being delayed by a counter traffic so large or so frequent as to interfere with the supply of post-horses.

At the end of each stage, which is generally only five miles, there is a large tile-roofed shed, built across the road, sufficiently high for a well-laden coach to pass under. The coachman pulls up to change horses beneath this, so that the traveller can get out of the carriage where he is protected from the heat of the sun. On each side of the shed are good stables, with raised brick and mortar flooring, and the neat cottage of the Native in charge of the posting-station, to whom all complaints are to be made, and who is bound to render immediate assistance. The horses stand ready harnessed, under the shed, at the time of your expected arrival, with their accoutrements in perfect order, all blacked and polished, and in excellent repair. The change of horses is effected nearly as quickly as it used to be in a fast coach in the best of the old coaching days, and unless the traveller wishes to alight for refreshment, the journey is continued without much more than a minute's interruption.

'At the end of each stage ... there is a large tile-roofed shed, built across the road....' (From Eliza Ruhamah Scidmore, *Java: The Garden of the East*, New York: Century Company, 1897)

We used frequently to avail ourselves of the opportunity of getting out under this large shed, and while they were changing horses, we made a most enjoyable breakfast. The large wooden table which stands at one side was soon decked by the willing natives with the large, clean, freshly-cut leaves of the plantain tree. On these the fresh bread, the hard-boiled eggs, the cold chicken, and the bottle of claret, which we had brought from our last night's resting-place, made a tempting display. The post-master's cottage supplied milk and hot water for the tea, and generally also fresh eggs for an omelet. These, and a most commendable Lyons sausage, which had accompanied us all the way from Batavia, were rendered doubly acceptable by the improving health and growing appetite of our dear invalid. We used to return to our comfortable English barouche, and to emerge from the protection of the shed into the bright sunny landscape beyond, with grateful hearts and in a happy spirit. L'illustrissima Dona Carolina Maria de Pinto de Cruz, our Portuguese maid, was sent to sit with her mistress inside, and a cigar, in the covered rumble of the carriage, formed the pleasant but prosaic close to our poetic repast. . . .

If you travel by night, an excellent kind of torch throws a bright light over the carriage and horses, and over the road in front, without inconveniencing the traveller either by smell or by smoke. It is made of long thin slits of dry bamboo, tied up into a pole as thick as a strong man's arm, and about eight feet long. This torch burns brightly, and is kept alight by the quick passage through the air, being held high up by one of the grooms behind, who leans it over to one side to let the sparks and burning embers fall clear of the carriage.

As the whole hilly interior of the island is very broken, with deep gorges and rushing streams, the constant help of buffaloes, and of coolies, is required in the steep passes, and of ferry boats in crossing the mountain torrents. Admirable arrangements are made for the immediate supply of all such requirements, and generally for forwarding the traveller on his journey without delay. Every part of the road is in charge of a petty native official, answerable that everything is kept in constant preparation, and punishable for all shortcomings, while the whole is kept in a high state of efficiency by continuous European supervision, and by the frequent unannounced passage of the high European and native officials of the district. . . .

Roads.—The main roads through the whole length of the island, and across it in some places, were originally made, at the beginning of this century, by Marshal Daendels, who had learnt the importance of roads from Napoleon; and the other cross roads have been made at different periods since we restored Java to the Dutch in 1816. Along the chief lines of communication the roads are double, one for cattle, and one for horse and carriage traffic. The carriage roads are macadamized, and both the carriage and cattle roads are kept in excellent order. With the exception of the few short lines of railway in India, Java posting is the only civilized mode of land travelling in the East. . . .

The light swing of a good English barouche, the change of place and of scenery without effort or fatigue, and the delicious climate of the hilly interior, act like magic on an invalid. . . .

Tjiandjoer [Cianjur] Races.—During our journey in the interior, I associated with the Dutch officials, planters, and landed proprietors, as well as with Englishmen settled in Java, many of whom were then travelling along the same road as ourselves for the Tjiandjoer races. These we were fortunate enough to see, and were astonished to find horses, trained by the neighbouring regents and other Native chiefs, competing with the horses of Europeans with success. The Regent of Tjiandjoer and the Regent of Bandong, who had come to Tjiandjoer for the race week, each gave a cup; and they or some of their chiefs had horses entered for most of the races.

The evening after our arrival, we English were invited to meet the other Europeans and the Native chiefs at the Regent's, to draw for the running horses in a lottery. I found a large assembly of men at the Regent's house, which was comfortably furnished, and arranged in the manner of the European houses in Java. The Regents and the Europeans took their seats round a table at one end of the long room, and a number of Native chiefs were seated on chairs at the other end. The chiefs did not approach the end of the room where the Regents were, as that would have involved the necessity of going down on their knees, the only position in which any Native of inferior rank can approach a Regent; but the Europeans went among the chiefs at their end of the room, shaking hands, renewing old acquaintances, and talking and laughing in a friendly and cordial manner. I saw the Resident, the only man there of higher rank than the Regents, holding a long and lively confabulation with, apparently, one of the least exalted of the chiefs.

The whole scene was an instance of the genuineness and cordiality of the friendly intercourse between European and Native, with which I was much struck.

A lottery was drawn for each race, amid much laughing at the unfortunate drawers of blanks. The whole conversation was in Malay, which most of the Europeans there, even the English, spoke well, but which the Dutch seemed to speak like their mother tongue; jokes flew about in Malay, from Native to European, and *vice versa*, amid shouts of laughter. I was lucky enough to draw a horse in one of the races. After the lottery was over, the horses drawn were successively put up for sale by auction, and knocked down to the highest bidder, who paid the amount of the bid, minus a small discount for the race fund, to the drawer of the horse. As I could not understand the bidding, which was all carried on in Malay, I entrusted the sale of my horse to one of the Dutch gentlemen, who managed it for me so successfully, as to repay me for all the other lottery tickets which I had drawn blank.

It did not, however, need any polyglot acquirements to understand the state of feeling between Natives and Europeans, which was, to me, the most important part of the scene....

Bandong Stag Hunt.—We were fortunate also in seeing, at Bandong, one of the grand autumn stag hunts peculiar to the Preanger. Most of the Europeans in the neighbourhood, whether official or otherwise, joined in the sport, together with about five hundred mounted Natives, including the Regent himself, and almost every Native official of the regency, and large numbers of the peasantry. Many of the horses were of Arab or Australian blood, though the great majority of the Natives rode mere Java ponies, and all were ridden without stirrups, and either barebacked, or with a mere pad. Each man carried by his side the goluck, or Java knife, some two feet long, like a short sword.

We rode in detached groups over a large grassy plain, and whenever a deer, buck, or doe started out of the long grass, the nearest group rode at it. The running was taken up by every group the game came near, till it was caught and cut down, by the first man who could succeed in striking it across the back with his knife. The head and neck are the sporting perquisite of the man who cuts it down, the body belongs to the Regent as lord of the country and of the game. Every man in the field rode as fairly and independently as an

English farmer, regardless of any rank but the Regent's; European and Native jostled and hustled for the first cut, in good humour and without rudeness, but the Natives' horses were so good, and they rode so well and so boldly, that not a single European there got a head. Between 8 a.m. and noon we killed forty-nine deer of the large Scotch red-deer kind; but I was told that the number was much smaller than usual, and that, at these grand hunts, which occur periodically in September and October, after the grass has been burnt, there have been sometimes hundreds killed in a day.

We then adjourned to breakfast in a summer-house, on the top of a hill well out in the plain, where the European ladies had been watching the sport. The breakfast was given by the Regent, who himself presided, eating and drinking like the rest, and who formally offered to each European a share of the venison, which had been brought in by mounted buffaloes, employed to beat the more swampy and jungly parts of the ground.

The long lines of buffaloes beating the jungle, the gaudy dresses, and the brightly gilt mushroom-shaped hats, of the Natives galloping in different directions, with perhaps a dozen deer on foot at the same time, the deep blue sky, the glowing sunshine, and the bright tinted hills surrounding the plain, made altogether one of the prettiest scenes it has ever been my lot to witness.

State of the Country.—I made various expeditions to see the scenery and plantations, besides several shooting excursions after rhinoceros, wild cattle, and deer, all of which are numerous about Bandong, and thus I had many opportunities of seeing out-of-the-way parts of the country. These excursions, together with the general kindness and hospitality of both officials and planters, my being fortunately able to speak French and German, and the pains taken to answer my interminable questions, and to give me information, enabled me to form an opinion on the state of the country and of its inhabitants.

Europeans.—The European planters and landed proprietors are men of education and refinement, holding social positions by family and by fortune, equal to the Dutch civilians or to the Indian civil servants. Their plantations and estates show much careful management, with considerable outlay in improving the cultivation, and in adding to the material welfare of their tenants and work-people. The Europeans of all classes, who are relatively more numerous in Java than in India, speak the native languages fluently, and English, as well as Dutch, associate on friendly and equal terms with the high Native officials, and treat the other Natives with consideration and kindness.

Native Chiefs.—The Natives of rank are all chiefs, and either present or expectant salaried servants of the Dutch Government; but with the exception of the chiefs in the Preanger districts, and some very few in other parts of the island, none of them are landed proprietors. They are active, frank co-operators with the Europeans in business and in sport. The Regent was consulted about the shooting expeditions, and sent orders to the local chiefs to get beaters and trackers for us, and to furnish us with horses and buffaloes when required. I was astonished to find that it was not the Dutch officials, but really the Regent and his chiefs, who governed the regency, though under the supervision of the European authorities; and few points of contrast with India struck me more than the personally stirring habits of the chiefs, who, so far from being sunk in sloth and sensuality as in Raffles' time, are now actively engaged, with considerable energy, in officially governing the people under European superintendence. . . .

Peasantry.—The lower classes of Natives are fairly industrious, the Native cultivation is excellent, and the artificial works of terracing and irrigation are extensive and well looked after, while the people are cheerful, apparently happy, and the richest peasantry I have seen in any country but North America. Beggars, whether religious or from want, must be very scarce, for we did not see one during our whole stay in the island; the general prosperity, with the strong ties of family affection and mutual support, which are the most honourable characteristics of Indian races, leaving but few objects, in Java, for the large charity inculcated and practised by all good Mussulmans. . . . In Java, as far as I could see or learn, the Native looked to the European for help and for advice, and the intercourse between them was respectful on one side and kind on the other.

J. W. B. Money, *Java, or How to Manage a Colony*, London: Hurst and Blackett, 1861, Vol. 1, pp. 2–5, 10–13, 15–19, 24–7, 31–5, 39.

11
Albert S. Bickmore Visits a Sugar Plantation, 1868

While still in his twenties, American naturalist Albert S. Bickmore spent four years making collections in the Malay Archipelago and East Asia. Here is his brief account of Surabaya and environs, where Javanese villagers under government supervision rotated the cultivation of rice with the cultivation of sugar cane for the local mill.

THE streets of Surabaya are narrow compared to those of Batavia; but they are far better provided with shade-trees of different species, among which the tamarind, with its highly compound leaves, appears to be the favorite. Here, as in all the other chief cities of the archipelago, the dusty streets are usually sprinkled by coolies, who carry about two large watering-pots. In the centre of the city, on an open square, is the opera-house, a large, well-proportioned building, neatly painted and frescoed within. In the suburbs is the public garden, nicely laid out, and abounding in richly-flowering shrubs. . . .

The United States steamship Iroquois was then lying in the roads, and our consular agent at this port invited Captain Rodgers, our consul from Batavia, who was there on business, and myself, to take a ride with him out to a sugar-plantation that was under his care. In those hot countries it is the custom to start early on pleasure excursions, in order to avoid the scorching heat of the noonday sun. We were therefore astir at six. Our friend had obtained a large post-coach giving ample room for four persons, but, like all such carriages in Java, it was so heavy and clumsy that both the driver and a footman, who was perched up in a high box behind, had to constantly lash our four little ponies to keep them up to even a moderate rate of speed. Our ride of ten miles was over a well-graded road, beautifully shaded for most of the way with tamarind-trees. . . .

Our road that morning led over a low country, which was devoted wholly to rice and sugar-cane. Some of these rice-fields stretched away on either hand as far as the eye could see, and appeared as boundless as the ocean. Numbers of natives were scattered through these wide fields, selecting out the ripened blades, which their religion requires them to cut off *one by one*. It appears an endless task thus to gather in all the blades over a wide plain. These are clipped off near the top, and the rice in this state, with the hull still on, is called 'paddy'. The remaining part of the stalks is left in the fields to enrich the soil. After each crop the ground is spaded or dug up with a large hoe, or ploughed with a buffalo, and afterward harrowed with a huge rake; and to aid in breaking up the clods, water to the depth of four or five inches is let in. This is retained by dikes which cross the fields at right angles, dividing them up into little beds from fifty to one hundred feet square. The seed is sown thickly in small plats at the beginning of the rainy monsoon. When the plants are four or five inches high they are transferred to the larger beds, which are still kept overflowed for some time. They come to maturity about this time (June 14th), the first part of the eastern monsoon, or dry season. Such low lands that can be thus flooded are called *sawas*. Although the Javanese have built magnificent temples, they have never invented or adopted any apparatus that has come into common use for raising water for their rice-fields, not even the simple means employed by the ancient Egyptians along the hill, and which the slabs from the palaces at Nineveh show us were also used along the Euphrates.

Only one crop is usually taken from the soil each year, unless the fields can be readily irrigated. Manure is rarely or never used, and yet the *sawas* appear as fertile as ever. The sugar-cane, however, quickly exhausts the soil. . . . On this account only one-third of a plantation is devoted to its culture at any one time, the remaining two-thirds being planted with rice, for the sustenance of the natives that work on that plantation. These crops are kept rotating so that the same fields are liable to an extra drain from sugar-cane only once

A Java sugar *fabrik* with 'chimneys pouring out dense volumes of black smoke'. (From J. J. van Braam, *Vues de Java*, Amsterdam, 1842)

in three years. On each plantation is a village of Javanese, and several of these villages are under the immediate management of a *controleur*. It is his duty to see that a certain number of natives are at work every day, that they prepare the ground, and put in the seed at the proper season, and take due care of it till harvest-time.

The name of the plantation we were to see was 'Seroenie'. As we neared it, several long, low, white buildings came into view, and two or three high chimneys, pouring out dense volumes of black smoke. By the road was a dwelling-house, and the 'fabrik' was in the rear. The canes are cut in the field and bound into bundles, each containing twenty-five. They are then hauled to the factory in clumsy, two-wheeled carts called *pedatis*, with a yoke of *sapis* [cattle]. On this plantation alone there are two hundred such carts. The mode adopted here of obtaining the sugar from the cane is the same as in our country. It is partially clarified by pouring over it, while yet in the earthen pots in which it cools and crystallizes, a quantity of clay, mixed with water, to the consistency of cream. The water, filtering through, washes the crystals and makes the sugar, which up to this time is of a dark brown, almost as white as if it had been refined. This simple process is said to have been introduced by some one who noticed that wherever the birds stepped on the brown sugar with their muddy feet, in those places it became strangely white. After all the sugar has been obtained that is possible, the cheap and impure molasses that drains off is fermented with a small quantity of rice. Palm-wine is then added, and from this mixture is distilled the liquor known as '*arrack*', which consequently differs little from rum. It is considered, and no doubt rightly, the most destructive stimulant that can be placed in the human stomach, in these hot regions. From Java large quantities are shipped to the cold regions of Sweden and Norway, where, if it is as injurious, its manufacturers are, at least, not obliged to witness its poisonous effects.

After the sugar has been dried in the sun it is packed in large cylindrical baskets of bamboo, and is ready to be taken to market and shipped abroad.

Albert S. Bickmore, *Travels in the East Indian Archipelago*, London: John Murray, 1868, pp. 60, 64, 66–9.

12
Alfred Russel Wallace Declares Java 'The Finest Tropical Island in the World', 1869

'The main object of all my journeys', wrote Alfred Russel Wallace in his preface to The Malay Archipelago, *'was to obtain specimens of natural history. . . .' In this he succeeded eminently, bearing home with him after eight years of explorations in South-East Asia specimens of more than 125,000 mammals, reptiles, birds, shells, butterflies, beetles, and other insects. Wallace's written account of his explorations is full of the chase and of scientific findings and musings. But Wallace, like many other nineteenth-century naturalists, larded his accounts of discovery with many a personal observation and traveller's tale. He spent three and a half months in Java in 1861. Although here and there the collecting was disappointing, Wallace was greatly impressed with the fecundity and abundance of the island itself, and with the Dutch administration of it. The passages below provide a sense both of Wallace the naturalist and Wallace the traveller; they also show that he was very much a man of his time in viewing Java's indigenous people as 'semi-barbarous' and quite fortunate really to be the recipients of the 'modern civilization . . . now spreading over the land'.*

I spent three months and a half in Java, from July 18th to October 31st, 1861, and shall briefly describe my own movements, and my observations on the people and the natural history of the country. To all those who wish to understand how the Dutch now govern Java, and how it is that they are enabled to derive a large annual revenue from it, while the population increases, and the inhabitants are contented, I recommend the study of Mr Money's excellent and interesting work, *How to Manage a Colony.* The main facts and conclusions of that work I most heartily concur in, and I believe that the Dutch system is the very best that can be adopted, when a European nation conquers or otherwise acquires possession of a country inhabited by an industrious but semi-barbarous people. . . .

The mode of government now adopted in Java is to retain the whole series of native rulers, from the village chief up to princes, who, under the name of Regents, are the heads of districts about the size of a small English county. With each Regent is placed a Dutch Resident, or Assistant Resident, who is considered to be his 'elder brother', and whose 'orders' take the form of 'recommendations', which are however implicitly obeyed. Along with each Assistant Resident is a Controller, a kind of inspector of all the lower native rulers, who periodically visits every village in the district, examines the proceedings of the native courts, hears complaints against the head-men or other native chiefs, and superintends the Government plantations. This brings us to the 'culture system', which is the source of all the wealth the Dutch derive from Java, and is the subject of much abuse in this country because it is the reverse of 'free trade'. To understand its uses and beneficial effects, it is necessary first to sketch the common results of free European trade with uncivilized peoples.

Natives of tropical climates have few wants, and, when these are supplied, are disinclined to work for superfluities without some strong incitement. With such a people the introduction of any new or systematic cultivation is almost impossible, except by the despotic orders of chiefs whom they have been accustomed to obey, as children obey their parents. The free competition of European traders, however, introduces two powerful inducements to exertion. Spirits or opium is a temptation too strong for most savages to resist, and to obtain these he will sell whatever he has, and will work to get more. Another temptation he cannot resist, is goods on credit. The trader offers him gay cloths, knives, gongs, guns, and gunpowder, to be paid for by some crop perhaps not yet planted, or some product yet in the forest. He has not sufficient forethought to take only a moderate quantity, and not enough energy to work early and late in order to get out of debt; and the consequence is that he accumulates debt upon debt, and often remains for years, or for life, a debtor and almost a slave. This is a state of things which occurs very largely in every part of the world in which men of a superior race freely trade with men of a lower race. It extends trade no doubt for a time, but it demoralizes the native, checks true civilization, and does not lead to any permanent increase in the wealth of the country; so that the European government of such a country must be carried on at a loss.

The system introduced by the Dutch was to induce the people, through their chiefs, to give a portion of their time to the cultivation of coffee, sugar, and other valuable products. A fixed rate of wages—low indeed, but about equal to that of all places where European competition has not artificially raised it—was paid to the labourers engaged in clearing the ground and forming the plantations under Government superintendence. The produce is sold to the Government at a low fixed price. Out of the net profits a percentage goes to the chiefs, and the remainder is divided among the workmen. This surplus in good years is something considerable. On the whole, the people are well fed and decently clothed; and have acquired habits of steady industry and the art of scientific cultivation, which must be of service to them in the future....

Taking it as a whole, and surveying it from every point of view, Java is probably the very finest and most interesting tropical island in the world. It is not first in size, but it is more than 600 miles long, and from sixty to 120 miles wide, and in area is nearly equal to England; and it is undoubtedly the most fertile, the most productive, and the most populous island within the tropics. Its whole surface is magnificently varied with mountain and forest scenery. It possesses thirty-eight volcanic mountains, several of which rise to ten or twelve thousand feet high. Some of these are in constant activity, and one or other of them displays almost every phenomenon produced by the action of subterranean fires, except regular lava streams, which never occur in Java. The abundant moisture and tropical heat of the climate causes these mountains to be clothed with luxuriant vegetation, often to their very summits, while forests and plantations cover their lower slopes. The animal productions, especially the birds and insects, are beautiful and varied, and present many peculiar forms found nowhere else upon the globe. The soil throughout the island is exceedingly fertile, and all the productions of the tropics, together with many of the temperate zones, can be easily cultivated. Java too possesses a civilization, a history and antiquities of its own, of great interest. The Brahminical religion flourished in it from an epoch of unknown antiquity till about the year 1478, when that of Mahomet superseded it.

The former religion was accompanied by a civilization which has not been equalled by the conquerors; for, scattered through the country, especially in the eastern part of it, are found buried in lofty forests, temples, tombs, and statues of great beauty and grandeur; and the remains of extensive cities, where the tiger, the rhinoceros, and the wild bull now roam undisturbed. A modern civilization of another type is now spreading over the land. Good roads run through the country from end to end; European and native rulers work harmoniously together; and life and property are as well secured as in the best governed states of Europe. I believe, therefore, that Java may fairly claim to be the finest tropical island in the world, and equally interesting to the tourist seeking after new and beautiful scenes; to the naturalist who desires to examine the variety and beauty of tropical nature; or to the moralist and the politician who want to solve the problem of how man may be best governed under new and varied conditions.

The Dutch mail steamer brought me from Ternate to Sourabaya [Surabaya], the chief town and port in the eastern part of Java, and after a fortnight spent in packing up and sending off my last collections, I started on a short journey into the interior. Travelling in Java is very luxurious but very expensive, the only way being to hire or borrow a carriage, and then pay half-a-crown a mile for post-horses, which are changed at regular posts every six miles, and will carry you at the rate of ten miles an hour from one end of the island to the other. Bullock carts or coolies are required to carry all extra baggage. As this kind of travelling would not suit my means, I determined on making only a short journey to the district at the foot of Mount Arjuna, where I was told there were extensive forests, and where I hoped to be able to make some good collections. The country for many miles behind Sourabaya is perfectly flat and everywhere cultivated, being a delta or alluvial plain watered by many branching streams. Immediately around the town the evident signs of wealth and of an industrious population were very pleasing; but as we went on, the constant succession of open fields skirted by rows of bamboos, with here and there the white buildings and tall chimney of a sugar-mill, became monotonous. The roads run in straight lines for several miles at a stretch, and are bordered by rows of dusty tamarind-trees. At each mile there are little guard-houses, where a policeman is stationed; and there is a wooden gong, which by means of concerted signals may be made to convey information over the country with great rapidity. About every six or seven miles is the post-house, where the horses are changed as quickly as were those of the mail in the old coaching days in England.

I stopped at Modjo-kerto [Mojokerto], a small town about forty miles south of Sourabaya, and the nearest point on the high road to the district I wished to visit. I had a letter of introduction to Mr Ball, an Englishman long resident in Java and married to a Dutch lady, and he kindly invited me to stay with him till I could fix on a place to suit me. A Dutch Assistant Resident as well as a Regent or native Javanese prince lived here. The town was neat, and had a nice open grassy space like a village green, on which stood a magnificent fig-tree (allied to the Banyan of India, but more lofty), under whose shade a kind of market is continually held, and where the inhabitants meet together to lounge and chat. The day after my arrival, Mr Ball drove me over to the village of Modjo-agong [Mojoagung], where he was building a house and premises for the tobacco trade, which is carried on here by a system of native cultivation and advance purchase, somewhat similar to the indigo trade in British India. On our way we stayed to look at a fragment of the ruins of the ancient city of Modjo-pahit [Majapahit], consisting of two lofty brick masses, apparently the sides of a gateway. The extreme perfection and beauty of the brickwork astonished me. The bricks are exceedingly fine and hard, with sharp angles and true surfaces. They are laid with great exactness, without visible mortar or cement, yet somehow fastened together so that the joints are hardly perceptible, and sometimes the two surfaces coalesce in a most incomprehensible manner. Such admirable brickwork I have never seen before or since. There was no sculpture here, but abundance of bold projections and finely-worked mouldings. Traces of buildings exist for many miles in every direction, and almost every road and pathway shows a foundation of brickwork beneath it—the paved roads of the old city. In the house of the Waidono [*wedono*] or district chief at Modjo-agong, I saw a beautiful figure carved in high relief out of a block of lava, and which had been found buried in the ground near the village. On my expressing a wish to obtain some such specimen, Mr B. asked the chief for it, and much to my surprise he immediately gave it me. It represented the Hindoo goddess Durga, called in Java, Lora Jonggrang (the exalted virgin). She has eight arms, and stands on the back of a kneeling bull....

The specimen I had obtained was a small one, about two feet high, weighing perhaps a hundredweight; and the next day we had it conveyed to Modjo-kerto to await my return to Sourabaya. Having decided to stay some time at Wonosalem, on the lower slopes of the Arjuna Mountain, where I was informed I should find forest and plenty of game, I had first to obtain a recommendation from the Assistant Resident to the Regent, and then an order from the Regent to the Waidono; and when after a week's delay I arrived with my baggage and men at Modjo-agong, I found them all in the midst of a five days' feast, to celebrate the circumcision of the Waidono's younger brother and cousin, and had a small room in an outhouse given me to stay in. The courtyard and the great open reception-shed were full of natives coming and going and making preparations for a feast which was to take place at midnight, to which I was invited, but preferred going to bed. A native band, or Gamelang, was playing almost all the evening, and I had a good opportunity of seeing the instruments and musicians. The former are chiefly gongs of various sizes, arranged in sets of from eight to twelve, on low wooden frames. Each set is played by one performer with one or two drumsticks. There are also some very large gongs, played singly or in pairs, and taking the place of our drums and kettledrums. Other instruments are formed by broad metallic bars, supported on strings stretched across frames; and others again of strips of bamboo similarly placed and producing the highest notes. Besides these there were a flute and a curious two-stringed violin, requiring in all twenty-four performers. There was a conductor, who led off and regulated the time, and each performer took his part, coming in occasionally with a few bars so as to form a harmonious combination. The pieces played were long and complicated, and some of the players were mere boys, who took their parts with great precision. The general effect was very pleasing,

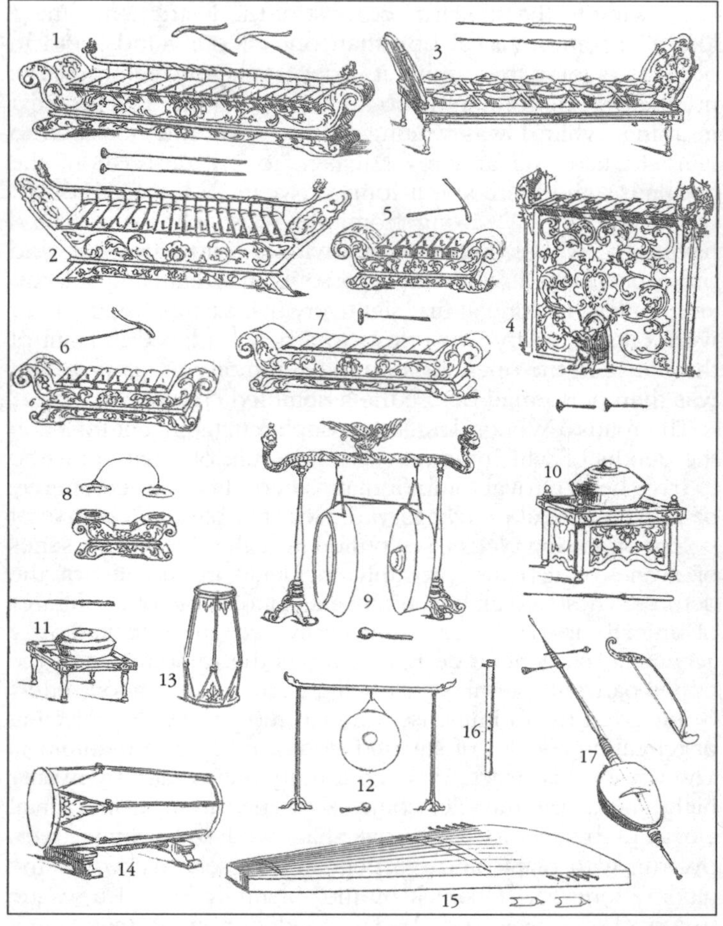

1	*Gambang Gangsa.*	5	*Saron.*	9	*Gong.*	14	*Kendang.*
2	*Gambang Kayu.*	6	*Demong.*	10	*Kenong.*	15	*Chelempung.*
3	*Bonang or Kroma.*	7	*Selantam.*	11	*Ketuk.*	16	*Suling.*
4	*Gender.*	8	*Kecher.*	12	*Kumpul.*	17	*Rebab.*
				13	*Ketipung.*		

Gamelan instruments, 'chiefly gongs of various sizes arranged ... on low wooden frames'. (From Thomas Stamford Raffles, *The History of Java*, London: Black, Parbury, and Allen, 1817) but, owing to the similarity of most of the instruments, more like a gigantic musical box than one of our bands; and in order to enjoy it thoroughly it is necessary to watch the large number of performers who are engaged in it. The next morning, while I was waiting for the men and horses who were to take me and my baggage to my destination, the two lads, who were about fourteen years old, were brought out, clothed in a sarong from the waist downwards, and having the whole body covered with a yellow powder, and profusely decked with white blossoms in wreaths, necklaces, and armlets, looking at first sight very like savage brides. They were conducted by two priests to a bench placed in front of the house in the open air, and the ceremony of circumcision was then performed before the assembled crowd.

The road to Wonosalem led through a magnificent forest, in the depths of which we passed a fine ruin of what appeared to have been a royal tomb or mausoleum. It is formed entirely of stone, and elaborately carved. Near the base is a course of boldly projecting blocks, sculptured in high relief, with a series of scenes which are probably incidents in the life of the defunct. These are all beautifully executed, some of the figures of animals in particular being easily recognizable and very accurate. The general design, as far as the ruined state of the upper part will permit of its being seen, is very good, effect being given by an immense number and variety of projecting or retreating courses of squared stones in place of mouldings. The size of this structure is about thirty feet square by twenty high, and as the traveller comes suddenly upon it on a small elevation by the roadside, overshadowed by gigantic trees, overrun with plants and creepers, and closely backed by the gloomy forest, he is struck by the solemnity and picturesque beauty of the scene, and is led to ponder on the strange law of progress, which looks so like retrogression, and which in so many distant parts of the world has exterminated or driven out a highly artistic and constructive race, to make room for one which, as far as we can judge, is very far its inferior.

Few Englishmen are aware of the number and beauty of the architectural remains in Java. They have never been popularly illustrated or described, and it will therefore take most persons by surprise to learn that they far surpass those of Central America, perhaps even those of India....

It is altogether contrary to the plan of this book to describe what I have not myself seen; but, having been led to mention them, I felt bound to do something to call attention to these marvellous works of art. One is overwhelmed by the contemplation of these innumerable sculptures, worked with delicacy and artistic feeling in a hard, intractable, trachytic rock, and all found in one tropical island. What could have been the state of society, what the amount of population, what the means of subsistence which rendered such gigantic works possible, will, perhaps, ever remain a mystery; and it is a wonderful example of the power of religious ideas in social life, that in the very country where, five hundred years ago, these grand works were being yearly executed, the inhabitants now only build rude houses of bamboo and thatch, and look upon these relics of their forefathers with ignorant amazement, as the undoubted productions of giants or of demons. It is much to be regretted that the Dutch Government do not take vigorous steps for the preservation of these ruins from the destroying agency of tropical vegetation; and for the collection of the fine sculptures which are everywhere scattered over the land.

Wonosalem is situated about a thousand feet above the sea, but unfortunately it is at a distance from the forest, and is surrounded by coffee-plantations, thickets of bamboo, and coarse grasses. It was too far to walk back daily to the forest, and in other directions I could find no collecting ground for insects. The place was, however, famous for peacocks, and my boy soon shot several of these magnificent birds, whose flesh we found to be tender, white, and delicate, and similar to that of a turkey. The Java peacock is a different species from that of India, the neck being covered with scale-like green feathers, and the crest of a different form; but the eyed train is equally large and equally beautiful. . . .

After a week at Wonosalem, I returned to the foot of the mountain, to a village named Djapannan [Japanan], which was surrounded by several patches of forest, and seemed altogether pretty well suited to my pursuits. The chief of the village had prepared two small bamboo rooms on one side of his own courtyard to accommodate me, and seemed inclined to assist me as much as he could. The weather was exceedingly hot and dry, no rain having fallen for several months, and there was, in consequence, a great scarcity of insects, and especially of beetles. I therefore devoted myself chiefly to obtaining a good set of the birds, and succeeded in making a tolerable collection. All the peacocks we had hitherto shot had had short or imperfect tails, but I now obtained two magnificent specimens more than seven feet long, one of which I preserved entire, while I kept the train only attached to the tail of two or three others. When this bird is seen feeding on the ground, it appears wonderful how it can rise into the air with such a long and cumbersome train of feathers. It does so, however, with great ease, by running quickly for a short distance, and then rising obliquely; and will fly over trees of a considerable height. I also obtained here a specimen of the rare green jungle-fowl (Gallus furcatus), whose back and neck are beautifully scaled with bronzy feathers, and whose smooth-edged oval comb is of a violet purple colour, changing to green at the base. It is also remarkable in possessing a single large wattle beneath its throat, brightly coloured in three patches of red, yellow, and blue. The common jungle-cock (Gallus bankiva) was also obtained here. It is almost exactly like a common gamecock, but the voice is different, being much shorter and more abrupt; whence its native name is Bekeko. Six different kinds of woodpeckers and four kingfishers were found here, the fine hornbill, Buceros lunatus, more than four feet long, and the pretty little lorikeet, Loriculus pusillus, scarcely more than as many inches.

One morning, as I was preparing and arranging my specimens, I was told there was to be a trial; and presently four or five men came in and squatted down on a mat under the audience-shed in the court. The chief then came in with his clerk, and sat down opposite them. Each spoke in turn, telling his own tale, and then I found out that those who first entered were the prisoner, accuser, policeman, and witness, and that the prisoner was indicated solely by having a loose piece of cord twined round his wrists, but not tied. It was a case of robbery, and after the evidence was given, and a few questions had been asked by the chief, the accused said a few words, and then sentence was pronounced, which was a fine. The parties then got up and walked away together, seeming quite friendly; and throughout there was nothing in the manner of any one present indicating passion or ill-feeling—a very good illustration of the Malayan type of character.

In a month's collecting at Wonosalem and Djapannan I accumulated ninety-eight species of birds, but a most miserable lot of insects. I then determined to leave East Java and try the more moist and luxuriant districts at the western extremity of the island. I returned to Sourabaya by water, in a roomy boat which brought myself, servants, and baggage at one-fifth the expense it had cost me to come to Modjo-kerto. The river has been rendered navigable by being carefully banked up, but with the usual effect of rendering the adjacent country liable occasionally to severe floods. An immense traffic passes down this river; and at a lock we passed through, a mile of laden boats were waiting two or three deep, which pass through in their turn six at a time.

A few days afterwards I went by steamer to Batavia, where I stayed about a week at the chief hotel, while I made arrangements for a trip into the interior....

The Hôtel des Indes was very comfortable, each visitor having a sitting-room and bedroom opening on a verandah, where he can take his morning coffee and afternoon tea. In the centre of the quadrangle is a building containing a number of marble baths always ready for use; and there is an excellent *table d'hôte* breakfast at ten, and dinner at six, for all which there is a moderate charge per day.

I went by coach to Buitenzorg, forty miles inland and about a thousand feet above the sea, celebrated for its delicious climate and its Botanical Gardens. With the latter I was somewhat disappointed. The walks were all of loose pebbles, making any lengthened wanderings about them very tiring and painful under a tropical sun. The gardens are no doubt wonderfully rich in tropical and especially in Malayan plants, but there is a great absence of skilful laying-out; there are not enough men to keep the place thoroughly in order, and the plants themselves are seldom to be compared for luxuriance and beauty to the same species grown in our hothouses.... Still, however, there is much to admire here. There are avenues of stately palms, and clumps of bamboos of perhaps fifty different kinds; and an endless variety of tropical shrubs and trees with strange and beautiful foliage. As a change from the excessive heats of Batavia, Buitenzorg is a delightful abode. It is just elevated enough to have deliciously cool evenings and nights, but not so much as to require any change of clothing; and to a person long resident in the hotter climate of the plains, the air is always fresh and pleasant, and admits of walking at almost any hour of the day. The vicinity is most picturesque and luxuriant, and the great volcano of Gunung-Salak, with its truncated and jagged summit, forms a characteristic background to many of the landscapes. A great mud eruption took place in 1699, since which date the mountain has been entirely inactive.

On leaving Buitenzorg, I had coolies to carry my baggage and a horse for myself, both to be changed every six or seven miles. The road rose gradually, and after the first stage the hills closed in a little on each side, forming a broad valley; and the temperature was so cool and agreeable, and the country so interesting, that I preferred walking. Native villages imbedded in fruit trees, and pretty villas inhabited by planters or retired Dutch officials, gave this district a very pleasing and civilized aspect; but what most attracted my attention was the system of terrace-cultivation, which is here universally adopted, and which is, I should think, hardly equalled in the world. The slopes of the main valley, and of its branches, were everywhere cut in terraces up to a considerable height, and when they wound round the recesses of the hills produced all the effect of magnificent amphitheatres. Hundreds of square miles of country are thus terraced, and convey a striking idea of the industry of the people and the antiquity of their civilization. These terraces are extended year by year as the population increases, by the inhabitants of each village working in concert under the direction of their chiefs; and it is perhaps by this system of village culture alone, that such extensive terracing and irrigation has been rendered possible.... The lower slopes of the mountains in Java possess such a delightful climate and luxuriant soil; living is so cheap and life and property are so secure, that a considerable number of Europeans who have been engaged in Government service, settle permanently in the country instead of returning to Europe. They are scattered everywhere throughout the more accessible parts of the island, and tend greatly to the gradual improvement of the native population, and to the continued peace and prosperity of the whole country.

Twenty miles beyond Buitenzorg the post road passes over the Megamendong Mountain, at an elevation of about 4,500 feet. The country is finely mountainous, and there is much virgin forest still left upon the hills, together with some of the oldest coffee-plantations in Java, where the plants have attained almost the dimensions of forest trees. About 500 feet below the summit level of the pass there is a road-keeper's hut, half of which I hired for a fortnight, as the country looked promising for making collections. I almost immediately found that the productions of West Java were remarkably different from those of the eastern part of the island; and that all the more remarkable and characteristic Javanese birds and insects were to be found here. On the very first day, my hunters obtained for me the elegant yellow and green trogon (Harpactes Reinwardti), the gorgeous little minivet flycatcher (Pericrocotus miniatus), which looks like a flame of fire as it flutters among the bushes, and the rare and curious black and crimson oriole (Analcipus sanguinolentus), all of them species which are found only in Java, and even seem to be confined to its western portion. In a week I obtained no less than twenty-four species of birds, which I had not found in the east of the island, and in a fortnight this number increased to forty species, almost all of which are peculiar to the Javanese fauna. Large and handsome butterflies were also tolerably abundant....

By far the most interesting incident in my visit to Java was a trip to the summit of the Pangerango [Pangrango] and Gedeh [Gede] mountains; the former an extinct volcanic cone about 10,000 feet high, the latter an active crater on a lower portion of the same mountain range. Tchipanas [Cipanas], about four miles over the Megamendong Pass [Puncak], is at the foot of the mountain. A small country house for the Governor-General and a branch of the Botanic Gardens are situated here, the keeper of which accommodated me with a bed for a night. There are many beautiful trees and shrubs planted here, and large quantities of European vegetables are grown for the Governor-General's table. By the side of a little torrent that bordered the garden, quantities of orchids were cultivated, attached to the trunks of trees, or suspended from the branches, forming an interesting open-air orchid-house. As I intended to stay two or three nights on the mountain I engaged two coolies to carry my baggage, and with my two hunters we started early the next morning. The first mile was over open country, which brought us to the forest that covers the whole mountain from a height of about 5,000 feet. The next mile or two was a tolerably steep ascent through a grand virgin forest, the trees being of great size, and the undergrowth consisting of fine herbaceous plants, tree-ferns, and shrubby vegetation. I was struck by the immense number of ferns that grew by the side of the road. Their variety seemed endless, and I was continually stopping to admire some new and interesting forms. I could now well understand what I had been told by the gardener, that 300 species had been found on this one mountain. A little before noon we reached the small plateau of Tjiburong [Cibeureum] at the foot of the steeper part of the mountain, where there is a plank-house for the accommodation of travellers. Close by is a picturesque waterfall and a curious cavern, which I had not time to explore. Continuing our ascent the road became narrow, rugged and steep, winding zigzag up the cone, which is covered with irregular masses of rock, and overgrown with a dense luxuriant but less lofty vegetation. We passed a torrent of water which is not much lower than the boiling point, and has a most singular appearance as it foams over its rugged bed, sending up clouds of steam, and often concealed by the overhanging herbage of ferns and lycopodia, which here thrive with more luxuriance than elsewhere.

At about 7,500 feet we came to another hut of open bamboos, at a place called Kandang Badak, or 'Rhinoceros-field', which we were going to make our temporary abode. Here was a small clearing, with abundance of tree-ferns and some young plantations of Cinchona. As there was now a thick mist and drizzling rain, I did not attempt to go on to the summit that evening, but made two visits to it during my stay, as well as one to the active crater of Gedeh. This is a vast semicircular chasm, bounded by black perpendicular walls of rock, and surrounded by miles of rugged scoria-covered slopes. The crater itself is not very deep. It exhibits patches of sulphur and variously-coloured volcanic products, and emits from several vents continual streams of smoke and vapour. The extinct cone of Pangerango was to me more interesting. The summit is an irregular undulating plain with a low bordering ridge, and one deep lateral chasm. Unfortunately there was perpetual mist and rain either above or below us all the time I was on the mountain; so that I never once saw the plain below, or had a glimpse of the magnificent view which in fine weather is to be obtained from its summit. Notwithstanding this drawback I enjoyed the excursion exceedingly, for it was the first time I had been high enough on a mountain near the Equator to watch the change from a tropical to a temperate flora. I will now briefly sketch these changes as I observed them in Java.

On ascending the mountain, we first met with temperate forms of herbaceous plants, so low as 3,000 feet, where strawberries and violets begin to grow, but the former are tasteless and the latter have very small and pale flowers. Weedy Compositæ also begin to give a European aspect to the wayside herbage. It is between 2,000 and 5,000 feet that the forests and ravines exhibit the utmost development of tropical luxuriance and beauty. The abundance of noble Tree-ferns, sometimes fifty feet high, contributes greatly to the general effect, since of all the forms of tropical vegetation they are certainly the most striking and beautiful. Some of the deep ravines which have been cleared of large timber are full of them from top to bottom; and where the road crosses one of these valleys, the view of their feathery crowns, in varied positions above and below the eye, offers a spectacle of picturesque beauty never to be forgotten. The splendid foliage of the broad-leaved Musacæ and Zingiberaceæ, with their curious and brilliant flowers, and the elegant and varied forms of plants allied to Begonia and Melastoma, continually attract the attention in this region. Filling up the spaces between the trees and larger plants, on every trunk and stump and branch, are hosts of Orchids, Ferns and Lycopods, which wave and hang and intertwine in ever-varying complexity. At about 5,000 feet I first saw horsetails (Equisetum), very like our own species. At 6,000 feet, Raspberries abound, and thence to the summit of the mountain there are three species of eatable Rubus. At 7,000 feet Cypresses appear, and the forest trees become reduced in size, and more covered with mosses and lichens. From this point upward these rapidly increase, so that the blocks of rock and scoria that form the mountain slope are completely hidden in a mossy vegetation. At about

8,000 feet European forms of plants become abundant. Several species of Honeysuckle, St. John's-wort, and Guelder-rose abound, and at about 9,000 feet we first meet with the rare and beautiful Royal Cowslip (Primula imperialis), which is said to be found nowhere else in the world but on this solitary mountain summit. It has a tall, stout stem, sometimes more than three feet high, the root leaves are eighteen inches long, and it bears several whorls of cowslip-like flowers, instead of a terminal cluster only. The forest trees, gnarled and dwarfed to the dimensions of bushes, reach up to the very rim of the old crater, but do not extend over the hollow on its summit. Here we find a good deal of open ground, with thickets of shrubby Artemisias and Gnaphaliums, like our southernwood and cudweed, but six or eight feet high; while Buttercups, Violets, Whortle-berries, Sow-thistles, Chickweed, white and yellow Cruciferæ, Plantain, and annual grasses everywhere abound. Where there are bushes and shrubs the St. John's-wort and Honeysuckle grow abundantly, while the Imperial Cowslip only exhibits its elegant blossoms under the damp shade of the thickets. . . .

In my more special pursuits, I had very little success upon the mountain, owing, perhaps, to the excessively unpropitious weather and the shortness of my stay. At from 7,000 to 8,000 feet elevation, I obtained one of the most lovely of the small fruit pigeons (Ptilonopus roseicollis), whose entire head and neck are of an exquisite rosy pink colour, contrasting finely with its otherwise green plumage; and on the very summit, feeding on the ground among the strawberries that have been planted there, I obtained a dull-coloured thrush, with the form and habits of a starling (Turdus fumidus). Insects were almost entirely absent, owing no doubt to the extreme dampness, and I did not get a single butterfly the whole trip; yet I feel sure that, during the dry season, a week's residence on this mountain would well repay the collector in every department of natural history.

After my return to Toego [Tugu], I endeavoured to find another locality to collect in, and removed to a coffee plantation some miles to the north, and tried in succession higher and lower stations on the mountain; but I never succeeded in obtaining insects in any abundance, and birds were far less plentiful than on the Megamendong Mountain. The weather now became more rainy than ever, and as the wet season seemed to have set in in earnest, I returned to Batavia, packed up and sent off my collections, and left by steamer on November 1st for Banca and Sumatra.

Alfred Russel Wallace, *The Malay Archipelago* . . ., London: Macmillan and Company, 1869, pp. 72–90, 92–3.

13
Henry O. Forbes Espies the Old and New Batavia and Observes Nature in West Java, 1879

Yet another naturalist, a Scot this time, Henry O. Forbes thoroughly took Java in during his 'wanderings in the Eastern Archipelago' between 1878 and 1883. In the following passages, he gives us his account of the old Batavia and the new—in which the Europeans commuted daily from the 'salubrious suburbs' by train—and describes a 'native market'. (This was soon de rigueur *for all travel writers.) At work in the field, Forbes discovers the astonishing accuracy of the indigenous classification system and observes with alarm the early warning signs of deforestation.*

BEFORE we had moored by the side of the Custom-house, it was quite dark, so that our landing was effected under some difficulty, amid the usual and necessary din and confusion, and amid a very Babel of foreign tongues, of which not a syllable was intelligible to me, save here and there a Portuguese word still recognisable, even after the changes of many centuries—veritable fossils bedded in the language of a race, where now no recollection or knowledge of the peoples who left them exists.

By dint of the universal language of signs, I got myself and baggage at last transferred to a carriage, drawn by two small splendidly running ponies, of a famous breed from the island of Sumbawa. After a drive of between two and three miles, through what seemed an endless row of Chinese bazaars and houses, remarkable mostly, as seen in the broken lamplight, for their squalor and stench, before which their occupants sat smoking and chatting, I at length emerged into a more genial atmosphere, and into canal and tree-margined streets, full of fine residences and hotels, very conspicuous by the blaze of light that lit up their pillared and marbled fronts.

Taking up my quarters at the Hotel der Nederlanden, I had to be content with an uncurtained shake-down on the floor of the room of one of my fellow passengers, as every bed in the hotel was occupied. Next morning, to every one's surprise, I arose without a single mosquito bite, evidently mosquito-proof. To my unspeakable comfort and advantage, I remained absolutely so during my whole sojourn in the East, and was thus relieved of the necessity of burdening myself with furniture against these, or any other insect pests whatever.

When the chaotic confusion of my first impressions of Batavia had become reduced to order, I found that it consisted of an old and a new town. The old town lies near the strand; is close, dusty, and stifling hot, standing scarcely anything above the sea-level. It contains the Stadthouse, the offices of the Government, with the various consulates and banks, all convenient to the wharf and the Custom-house, situated along the banks of canals, which intersect the town in every direction. Round this European nucleus cluster the native village, the Arab and the Chinese 'camps'.

Of Chinamen, Batavia contains many thousands of inhabitants, and, without this element, she might almost close her warehouses, and send the fleet that studs her roads to ride in other harbours; for every mercantile house is directly dependent on their trade. They are almost the sole purchasers of all the wares they have to dispose of. They rarely purchase except on credit, and a very sharp eye indeed has to be kept on them while their names are on the firm's books, for they are inveterate, but clever scoundrels, ever on the outlook for an opportunity to defraud. In every branch of trade, the Chinaman is absolutely indispensable, and, despite his entire lack of moral attributes, his scoundrelism and dangerous revolutionary tendencies, he must be commended for his sheer hard work, his indomitable energy and perseverance in them all. There is not a species of trade in the town, except, perhaps, that of bookseller and chemist, in which he does

In the Chinese 'camp'. (From M. T. H. Perelaer, *Het Kamerlid van Berkenstein in Nederlandsch-Indie*, Leiden, 1888–9)

not engage. Many of them possess large and elegantly fitted up *tokos* or shops, filled with the best European, Chinese, and Japanese stores; their workmanship is generally quite equal to European, and in every case they can far undersell their Western rivals.

The Arab, who like the Chinaman is prevented because of his intriguing disposition from going into the interior of the island, does, in a quiet and less obtrusive way, a little shop-keeping and money-lending, but is oftener owner of some sort of coasting craft, with which he trades from port to port, or to the outlying islands.

The natives of the town—that is, coast Malays and Sundanese—perform only the most menial work; they are vehicle drivers, the more intelligent are house servants, small traders, and assistants to the Chinese, but the bulk are coolies. They have no perseverance, and not much intelligence; and are very lazy, moderately dishonest, and inveterate gamblers, but otherwise innocuous.

This was the Batavia—fatal-climated Batavia—of past days. In this low-lying, close and stinking neighbourhood, devoid of wholesome water, scorched in the daytime, and chilled by the cold sea fogs in the night, did the Eastern merchant of half-a-century ago reside, as well as trade. Out of this, however, if he survived the incessant waves of fever, cholera, small-pox, and typhoid, he returned home in a few years, the rich partner of some large house, or the owner of a great fortune.

All this is changed now. Morning and evening, the train whirls in a few minutes the whole European population—which tries, in vain, to amass fortunes like those of past times—to and from the open salubrious suburbs, the new town, of fine be-gardened residences, each standing in a grove of trees flanking large parks, the greatest of which, the King's Plain, has each of its sides nearly a mile in length. Here the Governor-General has his official Palace—his unofficial residence being in the hills at Buitenzorg, about thirty miles to the south of Batavia; and here are built the barracks, the clubs, the hotels, and the best shops, dotted along roads shaded by leafy Hibiscus shrubs, or by the *Poinciana regia*, an imported Madagascar tree, which should be seen in the end of the year, when its broad spreading top is one mass of orange-red blossoms, whose falling petals redden the path, as if from the lurid glare of a fiery canopy above. To these pleasant avenues, in the cool of the evening, just after sunset, and before the dinner-hour, all classes, either driving or on foot resort for exercise and friendly intercourse. . . .

Mr Fraser's estate-house at Tjikandi-Udik [Cikandi Udik], which I reached late in the evening, I found to stand amid a rich and entirely cultivated country, but as regards my pursuits a barren territory. After enjoying for a few days the hospitality of the Administrator I moved south-westward to Genteng in the higher region of Lebak, where I was told some forest was then being felled.

Here I built a bamboo-hut in an open spot with an exhilarating look-out on the high mountains, and alone with my Malay boys began my initiation into the language of the country, and into the nomadic joyous life of a field naturalist. It is a life full of tiresome shifts, discomforts, and short commons; but these are completely forgotten, and the days seem never long enough amid that constant flash of delighted surprise that accompanies the beholding for the first time of beast or bird or thing unknown before, and the throb of pleasure experienced, as each new morsel of knowledge amalgamates with one's self.

Between myself and my boys for a time the most ludicrously comprehended sign-language was carried on, till their speech, whose sentences to my unaccustomed ears seemed composed of but one continuous word of innumerable uncouth syllables, began to shape itself into distinguishable elements, when to my amazement, as if some obstruction had been suddenly removed from my ears, I comprehended them as if I had been brought up among them. Before many weeks were over I could converse in the Malay tongue with an amount of freedom that surprised me.

The language of the district, that is, of the Sundanese themselves, though containing many Javanese and Malay words, is quite distinct from either. It is a coarser and rougher speech, and it was some time before I managed to acquire it; but I found it to be—like broad Scotch in comparison with pure English—one of great expressiveness.

As soon as I was able to follow their discourse with ease, my daily talks with these men were a source of great pleasure to me. I soon found out that in regard to every thing around them, they were marvellously observant and intelligent. Not one or two only, but every individual amongst them seemed equally stored with natural history information. There was not a single tree or plant or minute shrub, but they had a name for, and could tell the full history of; and not a note in the forest but they knew from what throat it proceeded. Every animal had a designation, not a mere meaningless designation, but a truly binomial appellation as fixed and distinctive as in our own system, differing only in the fact that theirs was in their own and not in a foreign language. Often enough this designation has so close a resemblance and sound to Latin, that it has been accepted by Western naturalists as if it had been so. One of the liveliest and most obtrusive of the squirrels in Java and Sumatra is a little red-furred creature called by the natives *tupai*, and to distinguish it from its more arboreal congeners they add, from its habit of frequenting branches near the ground, the word *tana* (for earth); and *Tupaia tana* is its accepted scientific term among European naturalists.

They have unconsciously classified the various allied groups into large comprehensive genera, in a way that shows an accuracy of observation that is astonishing from this dull-looking race. In this respect they excel far and away the rural population of our own country, among whom without exaggeration scarcely one man in a hundred is able to name one tree from another, or describe the colour of its flower or fruit, far less to name a tree from a portion indiscriminately shown him. How acute is their observation is exemplified by their name for the groups of true parasitic plants of the *Loranthaceæ* (or Misletoes), which are disseminated chiefly by being unobtrusively dropped by birds in convenient clefts of trees, they denominate as *Tai booroong* ('birds' excreta'); while to epiphytic plants they give a name that has almost the significance of our own scientific term. The great group of the Laurels, which so vary in flower and foliage as to be separated off into many genera by botanists, are all designated by the one name *Huru*, but they are differentiated by no fewer than sixty-three different specific terms, in every instance indicating some prominent distinguishing characteristic of flower, fruit or timber; and on examination, very few indeed of them turn out not to belong to the Laurel family. . . .

A stream which ran near my house was crossed by one of those native-made bamboo bridges, which spaciously housed and thatched over, have such a neat and attractive look about them. Every Sunday morning the district market was held under it, which from an early hour presented quite a gay and busy scene. I never missed, if I could, an opportunity of visiting these *Passars*, as I found them delightful resorts for studying the native in his gayer moods; for market-day was always their holiday, and the market-place the rendezvous for the youths and maidens of the district, as well as the news-exchange of the old men. The vendors, to be early at the market-place, generally spent Saturday evening and night under the shade of the bridge, or collected in the neighbouring village, whence the tinkle of the gamelang, their characteristic musical instrument, would be heard throughout the livelong night in company, if not concord, with the higher notes of their curiously drawling voices, repeating tjeritas [stories] or semi-historical tales, and adaptations from the Koran, varied by *pantuns* [short poems] or love songs.

The collection of wares exposed for barter was always a curious one; *sarongs* from their own looms—whose incessant click-clack is one of the most pleasant and characteristic of the industrial sounds in their villages—calicoes and silk kerchiefs from Manchester and Liverpool; Clark's Paisley thread of 'extra quality'; native-made horn combs, gay ornaments of spangles and beads, and the elaborate inlaid silver breast-pins for which the district is famous, worn by every female to fasten her loose upper robes; and bamboo hats in great variety. The Bantamese are specially noted for the manufacture of these last, and some of them are really exquisite specimens of plaiting. In the finest quality, made of carefully prepared narrow strips of the wood, a quiet but lucrative trade is done with European markets by unobtrusive go-betweens who collect them through the district. In Bantam they cost a mere trifle, but in Paris, I am informed, they are retailed at a profit of nearly one thousand per cent, as true Panama hats, from which it is difficult to distinguish them. One of these hats, that I treated to the roughest jungle work of three years, was scarcely impaired when we parted company.

Other than these the chief articles were household utensils, large copper jars for the preparation of rice, beat out of sheet copper by native smiths, and shallow iron basins (of Singapore make) for the daily extraction of the oil of the cocoa-nut palm, without which and its twin brother the bamboo, native prosperity and happiness would cease. There were besides piles of various species of dry-salted river fishes, chiefly Gabus (*Ophiocephalus striatus*), Soro and Regis (*Barbus duronensis* and *B. emarginatus*), and Gurame (*Ophromenus olfax*), the most prized of them all, in which a large and profitable trade is carried on with distant parts of the Archipelago. Many of these fishes are carefully preserved in the larger wet rice fields, where during the rainy season, having abundance of food, they multiply with great rapidity. During the hot season, when the *sawahs* have become, except in the centre, dry fields, the fishes are captured in immense numbers. Fried in fresh oil they form an excellent dish, and are the staple flesh-food of the natives.

A vile odour which permeates the whole air within a wide area of the market-place, is apt to be attributed to these piles of fish; but it really proceeds from another compound sold in round black balls, called *trassi*. My acquaintance with it was among my earliest experiences of house-keeping at Genteng. Having got up rather late one Sunday morning—an opportunity taken by one of my boys to go unknown to me to the market, which I had not then visited—I was discomfited by the terrific and unwonted odour of decomposition:—'My birds have begun to stink, confound it!' I exclaimed to myself. Hastily fetching down the box in which they were stored, I minutely examined and sniffed over every skin, giving myself in the process inflammation of the nostrils and eyes for a week after, from the amount of arsenical soap I inhaled; but all of them seemed in perfect condition. In the neighbouring jungle, though I diligently searched half the morning, I could find no dead carcase, and nothing in the 'kitchen-midden', where somehow I seemed nearer the source; but at last in the kitchen itself I ran it to ground in a compact parcel done up in a banana leaf.

'What on the face of creation is this?' I said to the cook, touching it gingerly.

'Oh! master, that is trassi.'

'Trassi? What is *trassi*, in the name of goodness!'

'Good for eating, master;—in stew.'

'Have *I* been eating it?'

'Certainly, master; it is *most* excellent (*enak sekali*).'

'You born fool! Do you wish to poison me and to die yourself?'

'May I have a goitre (*daik gondok*), master, but it *is* excellent!' he asseverated, taking hold of the foreskin of his throat, by the same token that a countryman at home would swear, '*As sure's Death!*'

Notwithstanding these vehement assurances, I made it disappear in the depths of the jungle, to the horror of the boy, who looked wistfully after it, and would have fetched it back, had I not threatened him with the direst penalties if I discovered any such putridity in my house again. I had then to learn that in every dish, native or European, that I had eaten since my arrival in the East, this Extract of Decomposition was mixed as a spice, and it would have been difficult to convince myself that I would come by-and-by knowingly to eat it daily without the slightest abhorrence. . . .

One of the most terrible scourges of the island, and for which no remedy seems possible, is the spread everywhere of a species of tall, slender cane—useless for fodder and good only for thatch—which the natives call alang-alang. Every spot unoccupied by forest, falls a prey to it; and when once it gets the upper hand, forest seeds refuse to root in it. Neither the incessant rains, nor the driest droughts of summer kill it. The fire may sweep the surface bare, it fails to touch the roots, which spring again in fresher vigour through the ashes. Deep shade alone seems to check its growth. The native in the hill regions does not make *sawahs* (which are good from year to year), but constantly takes in his fields by felling, where he lists, in the unbroken forest. As, after reaping for only two seasons this new land (on which he scatters his seed between the fallen trunks), he deserts it for a newer patch, broad tracts of the island are every year becoming covered with this ineradicable exhauster of the soil. . . .

Henry O. Forbes, *A Naturalist's Wanderings in the Eastern Archipelago . . . from 1878 to 1883*, London: Sampson, Low, Marston, Searle and Rivington, 1885, pp. 6–8, 53–5, 59–62.

14
John Whitehead Ascends Mount Bromo, 1880s

For the more adventurous travellers, there were always Java's live volcanoes. Mount Bromo in east Java is among the more spectacular, offering views of Java's highest mountain, Semeru, and, in the dry season from April to November (modern traveller Bill Dalton tells us), 'blood-red' sunrises.[1] *British ornithologist John Whitehead made the trip in the 1880s, beginning his ascent from a favourite Dutch hill station, Tosari.*

EARLY in September I made an excursion, accompanied by a young Dutchman, to the crater of the Bromo. After reaching the top of the spur on which Tosari is situated, the bridle-path runs for some distance along the ridge, then gradually slopes into a valley covered with a forest of Casuarina pines, rising again until the rim of the crater is reached, when by a steep zigzag path the sandy desert is gained. We started from the hotel at 5.30 a.m., mounted on ponies, and slowly ascended the Tosari spur; when the top of this is reached a wonderful panorama of distant hills and valleys unfolds itself to the view, but towering above all is one of the finest sights in Java, the cone of the Smeroe [Semeru], an active volcano of 12,000 feet, which in the clear morning air looks but a short distance off.

The summit of this volcano is bare some distance down, before the forest belt begins, forming a black ring round the pinkish-coloured conical top. After standing expectant some time—waiting for the next eruption—slowly a small ball issues from the point of the cone, which gradually becomes a rolling pillar of pinkish smoke; the pillar slowly rises and unrolls itself, gradually growing, and wonderfully spreads itself out until clear of the crater—rising and still spreading until it has become a simple cloud, to be dispersed by the winds in the higher air, and to fall in fine powder over the

[1]Bill Dalton, *Indonesia Handbook*, Chico, California: Moon Publications, 1991, p. 362.

'Early in September I made an excursion ... to the crater of the Bromo.' (From *Handbook of the Netherlands East Indies*, Buitenzorg: Department of Commerce, 1924)

surrounding country; but slowly as this little pillar seems to ascend in the distance, it is leaving the brink of that mighty crater with a velocity and roar greater than mere words can express, accompanied by huge masses of rock and ash, which fall again into the crater's mouth, securely blocking it up, to be blown skywards again when sufficient steam-power has been accumulated. This explosion takes place at intervals of a quarter of an hour. After standing for some while struck with the beauty of this scene, we rode on towards the crater of the Bromo, which lay to our left. On the path were tracks of Peafowl, one of which we flushed, but it disappeared too quickly for me to obtain a shot. After a few miles we arrived at the edge of the ancient crater of the Bromo, and from this point of vantage one of the most wonderful sights in the world may be contemplated. A thousand feet below you is a great level sandy desert, surrounded on all sides by barren mountains of volcanic ash, scarred and deeply furrowed in all directions; here and there numerous small craters have burst up, piling up huge fantastically formed mounds of ash: one of these mounds, several hundred feet high, resembles a gigantic cooking-mould, fluted all round with deep gullies equidistant from each other; this crater is called the Batok. The curious colouring of the landscape adds much to the wonder of the scene, the walls of the crater being of various shades of grey, the numberless deep scars on its surface casting shadows from blue to deep purple in the early morning sunlight, the atmosphere being so wonderfully clear at this time that the smallest irregularities of outline are distinctly visible. After descending the steep side of the crater we traversed the sandy desert, covered sparsely with patches of rank grass, the only animal life I noticed being a Pipit (*Anthus rufulus*). After a ride of some considerable distance across the desert, the base of a long hill composed entirely of soft ash is reached; this desert was at one time the active crater of the Bromo, but is now as it were roofed over by an even plain of lava, which is hollow beneath, the ground echoing with every footfall. At the base of the hill mentioned the ponies were left; the ascent of this hill is greatly facilitated by steps and hand-rails. On the way up steaming fissures in which sulphur has bubbled out may be noticed. This low hill of ash is now the only active crater of the Bromo, which in remote ages must have had an active crater of perhaps some twenty or twenty-five miles in circumference.

On reaching the rim of the crater, a narrow path leads almost round it, from which you can look down into the depths of a pit, the bottom of which is obscured by the ever-rising steam, which rolls over and over, seldom rising to the summit of the crater, accompanied by a dull roar like the boiling of a huge caldron; but to-day is a quiet day with the Bromo, for goodly-sized blocks of lava, strewed for hundreds of yards around on the sandy plain, testify to the fearful force at times exerted by this apparently harmless giant. It is only by seeing such sights that one is able to form an idea of the wonders of volcanic force, which at times have altered this world's surface over areas many miles in extent. A few months before I visited Tosari, the landlady of the hotel lost one of her sons, who was buried, together with an entire coffee-plantation, houses, occupants, and all, by a copious fall of boiling mud, which had been shot out of the Smeroe from a distance of several miles.

John Whitehead, *Exploration of Mount Kina Balu, North Borneo*, London: Gurney and Jackson, 1893, pp. 91–3.

15
David Fairchild Remembers the Java of His Youth, 1896

As a child growing up on the plains of Kansas, David Fairchild met the great Alfred Russel Wallace—who came to dinner one evening after having delivered a lecture at Kansas Agricultural College, of which Fairchild's father was President. Fairchild later studied plant science himself and, in 1896, at the age of twenty-seven and with Wallace's The Malay Archipelago *under his arm, made his way to Java. There he conducted research at Buitenzorg, where Melchior Treub was Director of the Botanical Gardens. Here are some excerpts from his memoirs of these happy days, written nearly half a century later.*

THE great day finally arrived and I sailed off for Java under Doctor Treub's chaperonage on a boat of the Netherlands-India Packet Boat Company.

That trip to Java in the nineties held more of interest than any ocean voyage I have ever taken. When I boarded the boat in Genoa and saw the turbaned Sundanese stewards, the Malays in their blue costumes, and the children's Javanese nurses, or 'baboos', in their sarongs, I realized that I was entering a new world. By the time we reached Port Saïd, I had acquired a considerable acquaintance on the boat.

Entering the Red Sea was a great event. When I came on deck the first morning 'east of Suez', the ladies lay in the chairs all along the deck clad in strange costumes or negligées—I was not sure which. They were nearly all barefooted and the sarongs which they wore were much shorter than the dresses of those days. I took a hasty look around and decided that I had mistaken the hour; that men were not supposed to be on deck so early. Turning abruptly, I fled from what seemed a definitely boudoir atmosphere. But Treub reassured me, explaining that the women had all donned native Javanese costume simultaneously because it was a fixed rule of the Captain that no lady could appear in native dress west of Suez....

Round-the-world tourists of today who find the streets of Batavia noisy with motor traffic, policemen and buses, can have no conception of the sleepy atmosphere which characterized that old Dutch town in the nineties. Its great open square, the Konings Plein, was surrounded by enormous Ficus trees and, in their shade, turbaned Javanese wandered barefoot, swinging their beautiful bamboo hats, or carrying on their shoulders long bamboo poles with baskets at each end. Tiny ponies trotted along, pulling some white-clad official sitting back to back with the driver in one of the two-wheeled carts called dos-à-dos.

In the hotel patio, the glare of midday was tempered by the shade of gigantic, overarching banyans, and an idyllic leisureliness pervaded everything everywhere. The 'boys', in beautiful, handmade sarongs of brown and indigo, came and went noiselessly across the patio and along the verandah, speaking softly to each other in a language as musical as the speech of an Andalusian. Every afternoon after luncheon, a hush des-

David Fairchild (centre) drinks from a fresh coconut, Java, 1896. (From David Fairchild, *The World Was My Garden*, New York: Charles Scribner's Sons, 1945)

cended during the siesta hour. We drowsed happily, listening to the chatter of small parrots in the branches of the trees and the occasional thud of a coconut as it fell to the ground.

The delightful village of Buitenzorg, where I spent eight months, lies in the saddle between two smoking volcanoes, the Salak and the Gedeh. A little railway train, manned by turbaned natives, took me there. Joyfully tooting its toy whistle, the train crossed the lowlands and wound up the mountainside through the most fairylike and utterly delightful scenery. The swampy plain was filled with giant ferns and Nipa palms, those stemless Oriental palms whose fronds resemble giant cycad leaves. A strange mist hangs over these lowlands, making it an unhealthy place to live.

Soon the hills were reached, with their kampongs composed of pretty, little bamboo houses thatched with palm leaves and shaded by lofty clumps of feathery bamboo gently waving in the breeze. Clean-swept pathways led to the springs or brooks, where naked children and their mothers were taking a morning bath. Surrounding every bamboo house was a bamboo woven hedge, each one a different pattern. Hanging in little bamboo baskets from bamboo poles were cooing doves, the favorite song birds of the Javanese. It seemed a civilization dependent upon bamboo for both necessities and luxuries.

Before noon, I was installed at the one-story Hotel Chemin de Fer in Buitenzorg, and here I lived during the eight months of my sojourn. When I opened the simple wooden shutters of my room, I looked out on a thoroughfare crowded with traffic, but yet a noiseless one. For although a human being passed my window almost every second throughout the day, he passed silently, barefooted.

My broad, hard bed was spotless, and on it lay one of those curious, long, round cushions with which the sleeper is supposed to separate his knees at night for coolness. Festooned above the bed, on a metal frame, was a mosquito net, another object which I had never seen before. Although this was in the days when few men besides Ross and Theobold Smith had an idea that the mosquito could be anything but a nuisance, yet there was an undefined suspicion that the miasma—the intangible something in the air which produced malaria—was kept out at night by the meshes of a mosquito netting.

Director Treub kindly insisted upon accompanying me when I made my first visit to the Gardens. I must have delighted him with my astonishment when he showed me the great avenue of huge Canarium trees (*Canarium commune*, Java Almond) which are so dramatically tropical with their weirdly buttressed roots. The trunk and limbs of the great trees were festooned with variegated climbing aroids (*Pothos aureus* and others) over which the liana, *Entada scandens*, had climbed like some gigantic reptile. At that time, this was undoubtedly the most remarkable avenue of trees in all the world.

Doctor Treub next showed me an orchid with a thousand blossoms (*Grammatophyllum speciosum*), and then took me to a tree from Africa which bore clusters of gold-edged, scarlet flowers. They were the great, flaming, cup-shaped blossoms of the African Tulip tree (*Spathodea campanulata*). As he picked up a fallen flower, he explained that, while in the bud, the gorgeous, red corolla had been held in its envelope of sepals by tough leathery bracts which contained a watery fluid under pressure. He had seen birds peck at these buds and had watched the fluid squirt from them. He thought the birds were frightened by the squirting fluid....

Director Treub next took me into the tangled jungle of the rattan Palms. I brushed against a swaying tip and instantly felt its grip upon my shoulder. Treub stood smiling at me as I struggled to free myself. The strength and sharpness of the clawlike spines was in surprising contrast to the delicacy and harmless appearance of the leaf tip. This tip is a climbing organ, quite as delicate as the tendril of the cucumber and other such vines.

The Director led me to the little laboratory which was placed at the disposal of visiting scientists. It was a great moment when I sat at the table designated as mine and realized that I had come to stay....

Life in the delightful laboratory was equalled in interest by life in the hotel. After a fatiguing morning in the Garden, I used to swing off the step of the dos-à-dos just in time for the 'rijsttafel' at the long hotel table. At this amazing meal we piled our deep plates high with steaming rice, and, as the long line of turbaned waiters offered them, added in turn a bit of fried chicken, slices of egg, perhaps a sardine, a meat-ball, or a fried banana. Then came the sambalang—a great tray containing red Macassar fish, tiny pickled ears of corn, burnt peanuts, roasted coconut, sweet mango chutney, a darker brand of Indian chutney as hot as liquid fire, a peculiarly flavored pickle made from Gnetum (a strange climbing shrub), and others which I cannot now recall. After making our choice, we poured over the heaping mass a quantity of curry sauce made fresh each day from ground-up cardamons and fiery-hot red peppers. The first mouthful of this mixture brought the perspiration to one's face and started something, whether it was digestion or not I do not know. However, in the end, one experienced a sense of well-being and drowsiness which admirably suited the tropical custom of a siesta after luncheon, and the hotel became as quiet as the grave from two to four....

My fellow boarders at the Hotel Chemin de Fer were a strange but fascinating lot. My next-door neighbor on one side drank innumerable 'bitterjes' [gin and bitters] and pestered me with strange inquiries about the English language, which he thought he spoke quite well.

'I am to be to are to want to go to bed. That is it English?' he once said.

My neighbor on the other side was a Belgian botanist named Cleautreau from Erera's laboratory in Brussels; a brilliant, emotional young man who wrote letters of forty pages in fine handwriting to his mother, and later, poor boy, died of a broken heart soon after his mother passed away.

DeMunnick, another hotel guest, had been a Dutch fonctionnaire. He was interested in schemes for utilizing the fibers of the Kapok or Silk-cotton tree which were then allowed to go to waste. He complained bitterly that his countrymen invested their money in American railroad stocks, but had nothing for the development of the Dutch East Indies.

This small group was augmented one morning by the arrival of Herr Rapps.... With his advent, our peaceful existence came to an end. DeMunnick soon confided to me that Rapps' baggage consisted of two enormous cases of Marsala wine and innumerable white sailor-caps. It also transpired that he had come to collect, not plants, but reptiles for a brother in Berlin.

Rapps sent word to the native quarter that he would buy live snakes and lizards, and swarms of natives appeared bearing all manner of creeping things. He soon had a dozen great, long lizards which he fastened to the legs of his sofa. When their activities kept him awake at night, he hung them out of the window, but this manœuvre was not successful, for they soon scratched the plaster off the walls and he had to cut them loose. The natives would catch them and sell them to him again the following morning. It was a mad performance from first to last.

Rapps also bought every available species of snake. We warned him about the deadly ular blang which, like the coral snake in Florida, has a mimic, a snake so nearly like it that it takes an expert to tell them apart.

I went into his disorderly bedroom one afternoon and saw an ular blang in a big candy jar with a cigarette stub holding the stopper open to give it air. I told Rapps that this was the poisonous snake, but he scoffed at me and said that he knew snakes, and this was the harmless mimic. Nevertheless I took the precaution of tying down the stopper before I left that night and made him promise that he would test the snake the next day.

In the morning, therefore, Rapps bought a young chicken. When I arrived, he had the snake tied loosely to a stanchion on the verandah and the chicken held in his bare hand as he presented the bird for the snake to strike at it. The snake most certainly would have struck Rapps' knuckles had I not pulled him back and helped him to arrange a safer test. I put the chicken in a box, covered it with a screen, and then put the snake in too. In a few minutes the chicken was dead. Needless to say, Rapps' reputation as a herpetologist sank rapidly to zero.

His behavior also went from bad to worse. He disgraced himself at Treub's immaculate retreat in the mountains by killing a wild boar, dressing it in the laboratory, and trying to make salt pork of it in a leaky flour barrel.

I cannot help looking back to those early days spent in Java with a longing which amounts to nostalgia, for the place which seemed a fairyland to a youth in his twenties has acquired a halo to the man in his seventies.

To a young man interested in nature it was then about the loveliest spot on the entire globe—a fairyland in which the quiet-voiced Javanese came and went softly. You did not hear their bare feet on the roadways; their batik costumes were the colors of autumn leaves that faded into the landscape; their meals of rice and fish and many kinds of vegetables were eaten noiselessly from wooden bowls, without a clatter of china; their thatched houses of bamboo seemed like playthings scattered picturesquely under the palms and mangosteen trees in the tiny kampongs; and the voices of children playing with crickets on the clean-swept dooryards mingled with the cooing of the turtle doves in bamboo cages which hung from the overarching tips of bamboo poles beside the eaves. . . .

These memories crowd in as I sit here to tell you what that island of Java was to a youth almost half a century back. I can bring back the pictures, but, alas, that Java disappeared long years ago.

I do not say that these things have vanished, but I do feel that wherever Western ways have penetrated they have carried something that disturbs the picturesqueness and quiet of the Javanese civilization, bringing into it crude elements of the machine-made world of which Westerners are proud. How could it be otherwise when great, burly, strong-armed, white-faced people who believe in big cities, shoved back the trees, brushed aside the tiny habitations, and straightened the winding roads?. . . .

The world is growing smaller, but I could not realize then what this shrinking of the world would do to such a paradise. It all seemed so permanent and so thoroughly lovely and quiet, this Java of mine in the nineties. . . .

David Fairchild, *The World Was My Garden*, New York: Charles Scribner's Sons, 1945, pp. 61–4, 71, 75–6. *Garden Islands of the Great East*, New York: Charles Scribner's Sons, 1943, pp. 6–8.

16

Eliza Ruhamah Scidmore Declares Java 'Finished', 1897

Eliza Ruhamah Scidmore, a prolific American traveller and travel writer of the late nineteenth and early twentieth centuries, had already chronicled her impressions of Alaska and Japan before making her trip to Java in the 1890s. 'All Java', she wrote, 'is in a way as finished as little Holland itself. . . . All the valleys, plains, and hillsides are planted in formal rows, hedged, terraced, banked, drained, and carefully weeded as a flower-bed.'[1] *To Scidmore, Java's villages appeared 'ornamental' and village life 'so prettily picturesque . . . that it seems only a theatrical representation'.*[2] *Scidmore made the Grand Tour, crossing Java from Batavia to Surabaya and taking full advantage of the island's new railway system. In the passages below, she comments on Java's rulers and subjects, on its trains, hotels, and plantations; and she takes in some of the island's famous sights: the Botanical Gardens at Buitenzorg, Borobudur, and the royal capital at Solo.*

THE Dutch do not welcome tourists, nor encourage one to visit their paradise of the Indies. Too many travelers have come, seen, and gone away to tell disagreeable truths about Dutch methods and rule; to expose the source and means of the profitable returns of twenty million dollars and more for each of so many years of the last and the preceding century—all from islands whose whole area only equals that of the State of New York. Although the tyrannic rule and the 'culture system', or forced labor, are things of the dark past, the Dutch official is still suspicious, and the idea being fixed fast that no stranger comes to Java on kindly or hospitable errands, the colonial authorities must know within twenty-four hours why one visits the Indies. They demand one's name, age, religion, nationality, place of nativity, and occupation, the name of the ship that brought the suspect to Java, and the name of its captain—a dim threat lurking in this latter query of holding the unlucky mariner responsible should his importation prove an expense or embarrassment to the island. Still another permit—a *toelatings-kaart*, or 'admission ticket'—must be obtained if one wishes to travel farther than Buitenzorg, the cooler capital, forty miles away in the hills. The tourist pure and simple, the sight-seer and pleasure traveler, is not yet quite comprehended, and his passports usually accredit him as traveling in the interior for 'scientific purposes'. Guides or efficient couriers in the real sense are all too rare. The English-speaking servant is rare and delusive, yet a necessity unless one speaks Dutch or Low Malay. Of all the countries

[1]Eliza Ruhamah Scidmore, *Java: The Garden of the East*, New York: Century Company, 1897, p. 71.

[2]Ibid., p. 143.

one may ever travel in, none equals Java in the difficulty of being understood; and it is a question, too, whether the Malays who do not know any English are harder to get along with than the Dutch who know a little....

The philippics of returned travelers furnish steady amusement for Singapore residents; and no one brings back the same enthusiasm that embarked with him. It is not the Java of the Javanese that these returned ones berate so vehemently, but the Netherlands India officials who impose so many hampering customs and restrictions upon all alien visitors and residents.... Java undoubtedly is 'the very finest and most interesting tropical island in the world', and the Javanese the most gentle, attractive, and innately refined people of the East, after the Japanese; but the Dutch in Java 'beat the Dutch' in Europe ten points to one, and there is nothing so surprising and amazing, in all man's proper study of mankind, as this equatorial Hollander transplanted from the cold fens of Europe; nor is anything so strange as the effect of a high temperature on Low-Country temperament. The most rigid, conventional, narrow, thrifty, prudish, and Protestant people in Europe bloom out in the forcing-house of the tropics into strange laxity, and one does not know the Hollanders until one sees them in this 'summer land of the world'....

Society [in Batavia] is naturally narrow, provincial, colonial, conservative, and insular, even to a degree beyond that known in Holland. The governor-general, whose salary is twice that of the President of the United States, lives in a palace at Buitenzorg, forty miles away in the hills, with a second palace still higher up in the mountains, and comes to the Batavia palace only on state occasions. This ruler of thirty odd million souls, who rules as a viceroy instructed from The Hague, with the aid of a secretary-general and a Council of the Indies, has, in addition to his salary of a hundred thousand dollars, an allowance of sixty thousand dollars a year for entertaining, and it is expected that he will maintain a considerable state and splendor....

The islands of Amboyna, Borneo, Celebes, and Sumatra are also ruled by this one governor-general of the Netherlands Indies, through residents; and the island of Java is divided into twenty-two residencies or provinces, a resident, or local

'The governor-general, whose salary is twice that of the President of the United States, lives in a palace at Buitenzorg.' (From M. T. H. Perelaer, *Het Kamerlid van Berkenstein in Nederlandsch-Indie*, Leiden, 1888–9)

governor, ruling—or, as 'elder brother', effectually advising—in the few provinces ostensibly ruled by native princes. A resident receives ten thousand dollars a year, with house provided and a liberal allowance made for the extra incidental expenses of the position—for traveling, entertaining, and acknowledging in degree the gifts of native princes. . . . All these residents are answerable to the secretary of the colony, appointed by the crown, and much of executive detail has to be submitted to the home government's approval. Naturally there is much friction between all these functionaries, and etiquette is punctilious to a degree. A formal court surrounds the governor-general, and is repeated in miniature at every residency. The pensioned native sovereigns, princes, and regents maintain all the forms, etiquette, and barbaric splendor of their old court life, elaborated by European customs. The three hundred Dutch officials condescend equally to the rich planters and to the native princes; the planters hate and deride the officials; the natives hate the Dutch of either class, and despise their own princes who are subservient to the Dutch; and the wars and jealousies of rank and race and caste, of white and brown, of native and imported folk, flourish with tropical luxuriance. . . .

In all the banks and business houses is found the lean-fingered Chinese comprador, or accountant, and the rattling buttons of his abacus, or counting-board, play the inevitable accompaniment to financial transactions, as everywhere else east of Colombo. The 563,449 Chinese in Netherlands India present a curious study in the possibilities of their race. Under the strong, tyrannical rule of the Dutch they thrive, show ambition to adopt Western ways, and approach more nearly to European standards than one could believe possible. Chinese conservatism yields first in costume and social manners; the pigtail shrinks to a mere symbolic wisp, and the well-to-do Batavian Chinese dresses faultlessly after the London model, wears spotless duck coat and trousers, patent-leather shoes, and, in top or derby hat, sits complacent in a handsome victoria drawn by imported horses, with liveried Javanese on the box. One meets correctly gotten-up Celestial equestrians trotting around Waterloo Plein or the alleys of Buitenzorg, each followed by an obsequious groom, the thin remnant of the Manchu queue slipped inside the coat being the only thing to suggest Chinese origin. The rich Chinese live in beautiful villas, in gorgeously decorated houses built on ideal tropical lines; and although having no political or social recognition in the land, entertain no intention of returning to China. They load their Malay wives with diamonds and jewels, and spend liberally for the education of their children. The Dutch tax, judge, punish, and hold them in the same regard as the natives, with whom they have intermarried for three centuries, until there is a large mixed class of these Paranaks [Peranakan] in every part of the island. . . .

The Javanese are the finer flowers of the Malay race—a people possessed of a civilization, arts, and literature in that golden period before the Mohammedan and European conquests. They have gentle voices, gentle manners, fine and expressive features, and are the one people of Asia besides the Japanese who have real charm and attraction for the alien. They are more winning, too, by contrast, after one has met the harsh, unlovely, and unwashed people of China, or the equally unwashed, cringing Hindu. They are a little people, and one feels the same indulgent, protective sense as toward the Japanese. Their language is soft and musical—'the Italian of the tropics'; their ideas are poetic; and their love of flowers and perfumes, of music and the dance, of heroic plays and of every emotional form of art, proves them as innately esthetic as their distant cousins, the Japanese, in whom there is so large an admixture of Malay stock. Their reverence for rank and age, and their elaborate etiquette and punctilious courtesy to one another, are as marked even in the common people as among the Japanese; but their abject, crouching humility before their Dutch employers, and the brutality of the latter to them, are a theme for sadder thinking, and calculated to make the blood boil. When one actually sees the quiet, inoffensive peddlers, who chiefly beseech with their eyes, furiously kicked out of the hotel courtyard when mynheer does not choose to buy, and native children actually lifted by an ear and hurled away from the vantage-point on the curbstone which a pajamaed Dutchman wishes for himself while some troops march by, one is content not to see or know any more....

The Weltevreden Station, on the vast Koenig's Plein, a spacious, stone-floored building, whose airy halls and waiting- and refreshment-rooms are repeated on almost as splendid scale at all the large towns of the island, was enlivened with groups of military officers, whose heavy broadcloth uniforms, trailing sabers, and clanking spurs transported us back from the tropics to some chilly European railway-station, and presented the extreme contrast of colonial life. The train that came panting from Tandjon Priok [Tanjungpriok] was made up of first, second, and third-class cars, all built on the American plan, in that they were long cars and not carriages, and we entered through doors at the end platforms. The first-class cars swung on easy springs; there were modern car-windows in tight frames, also window-frames of wire netting; while thick wooden venetians outside of all, and a double roof, protected as much as possible from the sun's heat. The deep arm-chair seats were upholstered with thick leather cushions, the walls were set with blue-and-white tiles repeating Mauve's and Mesdag's pictures, and adjustable tables, overhead racks, and a dressing-room furnished all the railroad comforts possible.... The second-class cars in Java rest on springs also, but more passengers are put in a compartment, and the fittings are simpler; while the open third-class cars, where native passengers are crowded together,

Passengers await the train at Tangung Station. (From M. T. H. Perelaer, *Het Kamerlid van Berkenstein in Nederlandsch-Indie*, Leiden, 1888–9)

have a continuous window along each side, and the benches are often without backs. . . .

Dutch engineers built and manage the road, but the staff, the working force of the line, are natives, or Chinese of the more or less mixed but educated class filling the middle ground between Europeans and natives, between the upper and lower ranks. Wonderful skill was shown in leading the road over the mountains, and in building a firm track and bridges through the reeking swamps, where no white man could labor, even if he could live. The trains do not run at night, which would be a great advantage in a hot country, for the reason that the train crews are composed entirely of natives (since such work is considered beneath the grade of any European), and the cautious Dutch will not trust native engineers after dark. . . .

The ordinary bedroom of a Java hotel, with latticed doors and windows, contains one or two beds, each seven feet square, hung with starched muslin curtains that effectually exclude the air, as well as lizards or winged things. The bedding, as at Singapore, consists of a hard mattress with a sheet drawn over it, a pair of hard pillows, and a long bolster laid down the middle as a cooling or dividing line. Blankets or other coverings are unneeded and unknown, but it takes one a little time to become acclimated to that order in the penetrating dampness of the dewy and reeking hours before dawn. If one makes protest enough, a thin blanket will be brought, but so camphorated and mildew-scented as to be insupportable. Pillows are not stuffed with feathers, but with the cooler, dry, elastic down of the straight-armed cotton-tree, which one sees growing everywhere along the highways, its rigid, right-angled branches inviting their use as the regulation telegraph-pole. The floors are made of a smooth, hard cement, which harbors no insects, and can be kept clean and cool. Pieces of coarse ratan matting are the only floor-coverings used, and give an agreeable contrast to the dirty felts, dhurries, and carpets, the patches of wool and cotton and matting, spread over the earth or wooden floors of the unspeakable hotels of British India. And yet the Javanese hotels are disappointing to those who know the solid comforts and immaculate order of certain favorite hostelries of The Hague and Amsterdam. Everything is done to secure a free circulation of air, as a room that is closed for a day gets a steamy, mildewed atmosphere, and if closed for three days blooms with green mold over every inch of its walls and floors. The section of portico outside each room at Buitenzorg was decently screened off to serve as a private sitting-room for each guest or family in the hours of startling dishabille; and as soon as the sun went down a big hanging-lamp assembled an entomological congress. Every hotel provides as a night-lamp for the bedroom a tumbler with an inch of cocoanut-oil, and a tiny tin and cork arrangement for floating a wick on its surface. For the twelve hours of pitch-darkness this little lighting-bug contrivance burns steadily, emitting a delicious nutty fragrance, and allowing one to watch the unpleasant shadows of the lizards running over the walls and bed-curtains, and to look for the larger, poisonous brown gecko, whose unpleasant voice calling '*Becky! Becky! Becky!*' in measured gasps, six times, over and over again, is the actual, material nightmare of the tropics. . . .

The famous Botanical Garden at Buitenzorg is the great show-place, the paradise and pride of the island. The Dutch are acknowledgedly the best horticulturists of Europe, and with the heat of a tropical sun, a daily shower, and nearly a century's well-directed efforts, they have made Buitenzorg's garden first of its kind in the world, despite the rival efforts of the French at Saigon, and of the British at Singapore, Ceylon, Calcutta, and Jamaica. The governor-general's palace, greatly enlarged from the first villa of 1744, is in the midst of the ninety-acre inclosure reached from the main gate, near the hotel and the passer [market], by what is undoubtedly the finest avenue of trees in the world. These graceful kanari-trees, arching one hundred feet overhead in a great green cathedral aisle, have tall, straight trunks covered with stag-horn ferns, bird's-nest ferns, ratans, creeping palms, blooming orchids, and every kind of parasite and air-plant the climate allows; and there is a fairy lake of lotus and *Victoria regia* beside it, with pandanus and red-stemmed Banka palms crowded in a great sheaf or bouquet on a tiny islet. When one rides through this green avenue in the dewy freshness of the early morning, it seems as though nature and the tropics could do no more, until he has penetrated the tunnels of waringen-trees, the open avenues of royal palms, the great plantation of a thousand palms, the grove of tree-fern, and the frangipani thicket, and has reached the knoll commanding a view of the double summit of Gedeh [Gede] and Pangerango [Pangrango], vaporous blue volcanic heights, from one peak of which a faint streamer of smoke perpetually floats. There is a broad lawn at the front of the palace, shaded with great waringen-, sausage-, and candle-trees, and trees whose branches are hidden in a mantle of vivid-leafed bougainvillea vines, with deer wandering and grouping themselves in as correct park pictures as if under branches of elm or oak, or beside the conventional ivied trunks of the North.... Over one hundred native gardeners tend and care for this great botanic museum of more than nine-thousand living specimens, all working under the direction of a white head-gardener....

All agree that, of all exiled cultivators in the far parts of the world, the Java planter is most to be envied, leading, as he does, the ideal tropical life, the one best worth living, in a land where over great areas it is always luxurious, dreamy afternoon, and in the beautiful hill-country is always the fresh, breezy, dewy summer forenoon of the rarest June.

The most favored and the most famous plantations are those around Buitenzorg and in the Preanger regencies, which lie on the other side of Gedeh and Salak, those two sleeping volcanoes that look down upon their own immediate foothills and valleys, to see those great, rolling tracts all cultivated like a Haarlem tulip-bed. Above the cacao limit, tea-gardens, coffee-estates, and kina-plantations [cinchona/quinine] cover all the land lying between the altitudes of two thousand and four thousand feet above the level of the sea. The owners of these choicest bits of 'the Garden of the East' lead an existence that all other planters of growing crops, and most people who value the creature comforts, the luxuries of life, and nature's opulence, may envy....

With the nearest neighbor ten miles away, and the thousand workmen employed upon the place settled with their families in different villages within its confines, Sinagar is a little world or industrial commune by itself, its master a patriarchal ruler, whose sway over these gentle, childlike Javanese is as absolute as it is kindly and just. The 'master' has sat under his Sinagar palms and gorgeous bougainvilleas for twenty-six out of the thirty-three years spent in Java, and his sons and daughters have grown up there, gone to Holland to finish their studies, and, returning, have made Sinagar a social center of this part of the Preangers. The life is like that of an English country house, with continental and tropical additions that unite in a social order replete with pleasure and interest. Weekly musicales are preceded by large dinner-parties, guests driving from twenty miles away and coming by train; and, with visitors in turn from all parts of the world, the guest-book is a polyglot and cosmopolitan record of great interest. Long wings have been added to the original bungalow dwelling, inclosing a spacious court, or garden, all connected by arcades and all illuminated by electric lights. The ladies' boudoir at the far end of the buildings opens from a great portico, or piazza, furnished with the hammocks, the ratan furniture, and the countless pillows of a European or American summer villa, but looking out on a marvelous flower-garden and an exquisite landscape view....

We strolled ... through a toy village under a kanari avenue, where all the avocations and industries of Javanese life were on view, and the little people, smiling their welcome, dropped on their heels in the permanent courtesy of the *dodok*, the squatting attitude of humility common to all Asiatics. The servants who had brought notes to the master, as he sat on the porch, crouched on their heels as they offered them, and remained in that position until dismissed; and the villagers and wayfarers, hastily dropping on their haunches, maintained that lowly, reverent attitude until we had passed—an attitude and a degree of deference not at all comfortable for an American to contemplate, ineradicable old Javanese custom as it may be....

The bachelor planter partner showed us his bungalow, full of hunting-trophies—skulls and skins of panthers, tigers, and wild dogs; tables made of rhinoceros-hide resting on rhinoceros and elephant skulls, and tables made of mammoth turtle-shells resting on deer-antlers. The great prizes were the nine huge banteng skulls, trophies of hunting-trips to the South Preanger, the lone region bordering on the Indian Ocean. There were also chandeliers of deer-antlers, and a frieze-like wall-bordering of python-skins, strange tusks and teeth, wings and feathers galore, and dozens of kodak pictures as witnesses and records of the many camps and battues of this sportsman—all gathered in that same wild region of big game, as much as fifty or a hundred miles away, but referred to in the Buitenzorg neighborhood as New York sportsmen used to speak of the buffalo country—'the south coast' and 'out West', equally synonyms for all untamed, far-away wildness....

With the younger people of the master's family, his young managers and assistants, fresh from Amsterdam schools and European universities, speaking English and several other languages, *au courant* with all the latest in the world's music, art, literature, and drama, plantation life and table-talk were full of interest and varied amusements. By a whir of the telephone, two of the assistants were bidden ride over from their far corner of the estate for dinner, and afterward a quartet of voices and instruments made the marble-floored music-room ring, while the elder men smoked meditatively, or clicked the billiard-balls in their deliberate, long-running tourney. The latest books and the familiar American magazines strewed boudoir and portico tables, and naturally there was talk of them.

'Ah, we like so much your American magazines—the "Century" and the others. We admire so much the pictures. And then all those stories of the early Dutch colonists at Manhattan!'....

They were undoubtedly disappointed that we did not speak Dutch, or at least read it, since all Hollanders know that Dutch is the language of the best families in New York, of the cultivated classes and all polite society in the United States, since from the mynheers of Manhattan came the first examples of refined living in the New World. 'The English colonists were of all sorts, you know, like in Australia,' said our informants at Buitenzorg and everywhere else on the island, 'and that is why you Americans are all so proud of your Dutch descent.'

Tea-bushes covered thousands of acres around and below us, as the ground dropped away from that commanding ridge, their formal rows decreasing in perspective until they shaded the landscape like a fine line-engraving. For mile after mile one could walk in direct line between soldierly files of tea-bushes—Chinese, Assam, and hybrids.... Each particular bush is tended and guarded as if it were the rarest ornamental exotic, and the tea-gardens, with their broad stripings of green upon the red ground, and skeleton lines of palms outlining the footpaths and the divisional limits of each garden, are like a formal exhibit of tea-growing, an exposition model on gigantic scale, a fancy farmer's experimental show-place....

The tea-pickers, mostly women, gather the leaves only when the plants are free from dew or rain. They pick with the lightest touch of thumb and finger, heaping the leaves on a square cloth spread on the ground, and then tying up the bundle and 'toting' it off on their heads, for all the world like the colored aunties of our southern states. The bright colors of their jackets and sarongs, and of their bundles, that look like exaggerated bandana turbans, give gay and picturesque relief to the green-striped gardens, whose exact lines converge in long, monotonous perspective whichever way one looks. There is great fascination in watching these bobbing figures among the bushes gradually converge to single lines, and the procession of lank, slender sarongs file through the gardens, down the avenues of palm and tamarind, to the fabrik. . . .

The Sundanese who live in their ornamental little fancy baskets of houses beneath Sinagar's tall tamarinds and kanari-trees are much to be envied by their people. The great estate is a world of its own, an agricultural Arcadia, where life goes on so happily that it is most appropriate that they should have presented model Javanese village life at the Chicago Exposition in 1893. These little Sinagar villagers have their frequent passers on one side or the other of the demesne by turn, with theater and *wayang-wayang*, or puppet-shows, lasting far into the night. Professional raconteurs thrill them with classic tales of their glorious past, while musicians make sweet, sad melodies to rise from *gamelan*, or *gambang kayu*, from fiddle, drum, bowls, bells, and the sonorous *angklung*—a rude instrument of most ancient origin, made of five or eight graduated bamboo tubes, cut like organ-pipes, and hung loosely in a frame, which, shaken by a master hand, or swinging in the breeze from some tree-branch, produces the strangest, most weird and fascinating melodies in all the East.

The play of village life about Sinagar is so prettily picturesque, so well presented and carried out, that it seems only a theatrical representation—a Petit Trianon sort of affair at the least. The smiling little women, who rub and toss tea-leaves over the wilting-trays at the fabrik, seem only to be playing with the loose leaves like a larger sort of intelligent, careful children. . . .

Beyond the region of the great plantations, where every hillside is cleared and planted up to the kina limit, and only the summits and steepest slopes are left to primeval jungle, there succeed great stretches of wild country, where remarkable engineering feats were required of the railway-builders. With two heavy engines the train climbs to Tjandjoer [Cianjur] station, sixteen hundred feet above the sea; and there, if one has telegraphed the order ahead, he may lunch at ease in his compartment as the train goes on. He may draw from the three-storied lunch-basket handed in either a substantial riz tavel, consisting of a little of everything heaped upon a day's ration of boiled rice, or a 'tiffin', whose *pièce de rèsistance* is a huge *bifstek mit ard appelen*, that would satisfy the cravings of any three dragoons. Either feast is followed by bread or bananas, with a generous section of a cheese, with mangosteens or other fruits, and one feels that he has surely reached the land of plenty and solid, solid comforts, where fate cannot harm him—when all this may be handed in to fleeting tourists at a florin and a half apiece.

After this station of abundant rations, all signs of cultivation and occupancy disappear, and the station buildings and the endless lantana-hedges along the railway-track are the only signs of human habitation or energy in the wilderness of hills covered with alang-alang or bamboo-grass, and the coarse *glagah* reeds which cattle will not touch.... The streams that come cascading down from all these green heights have carved out some beautiful scenery, and the Tjitaroem [Citarum] River, foaming in sight for a while, disappears, runs through a mountain by a natural tunnel, and reappears in a deep gorge, of which one has an all-too-exciting view as the train crosses on a spidery viaduct high in [the] air....

The train climbs very slowly from Bandong to Kalaidon Pass, and, after toiling with double engines up the steep grades, it rests at a level, and there bursts upon one the view of the plain of Leles—the fairest of all tropical landscapes, a vision of an ideal promised land, and such a dream of beauty that even the leaden blue clouds of a rainy afternoon could not dim its surpassing loveliness. The railway follows a long shelf hewn high on the mountain wall, that encircles an oval plain set with two conical mountains that rise more than two thousand feet above the level of this plain of Leles, itself two thousand feet above the level of the sea. The finely wrought surface of the plain—networked with the living green dikes and terraces of rice-fields, which, flooded, gleam and glitter in the fitful sun-rays, or, sown and harvested, glow with a mosaic of green and gold—is one exquisite symphony in color, an arrangement in greens that holds one breathless with delight....

The sesquipedalian names of the railway-stations throughout the Preanger regencies, are something to fill a traveler's mind between halts, and almost explain why the locomotives not only toot and whistle nearly all the time they are in motion, but stand on the track before station sign-boards and shriek for minutes at a time, like machines demented. Radjamendala is an easy arrangement in station names for the early hours of the trip, and all that family of names—Tjitjoeroek, Tjibeber, Tjirandjang [Cicurug, Cibeber, Ciranjang, etc.], Tjipenjeum, Tjitjalenka, and also Tagoogapoe—will slip from the tongue after a few trials; but when one strains his eyes toward the limits of the plain of Leles, he may almost see the houses of Baloeboer-Baloeboer-Limbangan [Bluburlimbangan]. People actually live there and pay taxes, and it is my one regret that I did not leave the train, drive over, and have some letters postmarked with that astonishing aggregation of sound-symbols. Only actual sight, too, could altogether convince one, that one small village of metal-workers could really support so much nomenclature together with any amount of profitable trade. In the intervals of practising the pronunciation of that particular geographic name, the artisans of Baloeboer-Baloeboer-Limbangan do hammer out serviceable gongs, bowls, and household utensils of brass and copper. In earlier times Baloeboer-Baloeboer-Limbangan was the Toledo of the isles, and the kris-blades forged there had finer edge than those from any other place in the archipelago. In these railroad and tramp-steamer days of universal, wholesale trade rivalry, the blade of the noble kris more often comes from abroad, and the chilled edges from Birmingham or those made in Germany have displaced the blades made at the edge of the plain of Leles, and the glory of Baloeboer-Baloeboer-Limbangan has departed....

In Middle Java, where the railway descends from the Preanger hills to the *terra ingrata*'s succession of jungle and swamp at the coast-level, one experiences the same dull, heavy, sickening, depressing heat as in Batavia. After the clear, fresh, mildly cool air, the eternal southern-California climate of the hills, this sweltering atmosphere gave full suggestion of the tropics' deadly perils. Hour after hour the train followed a raised embankment across an endless swamp, the brilliantly flowered lantana-hedges still accompanying the tracks, and a dense forest wall, tangled and matted together with ratans and other creepers, shutting off the view on either side. The malaria and the deadly fever-germs that haunt this region were almost visible, so dense was the air....

As we coursed along past those miles of rankest vegetation, not a waft of perfume reached us, nor did any mass of color or cloak of blossoms delight the eye—a green monotony of uninteresting vegetation, save for the ratan-palms which decorated every tree with their beautiful pinnate leaves.... No clouds, cascades, or festoons of gorgeous flowers, no waves of overpowering perfume, no masses of orchids, rewarded eager scrutiny; no birds of brilliant plumage flashed across the jungle's front; no splendidly striped tigers licked their chops and snarled in the jungle's shade; no rhinoceros snorted at the iron horse; and not a serpent raised a hissing head, slid away through dank grass, or looped itself from tree-top to tree-top in proper tropical fashion, as we steamed across the deadly, uninhabitable *terra ingrata.* Nor had even the first construction gangs of railway-builders met with any such sensational incidents, so the chief engineer of the railways afterward informed us....

'But there are great snakes in the swamps surely? You must run over them often?' we persisted.

'Doubtless; but we rarely see snakes here in Java. There are many in Borneo, Sumatra, and the other islands that are so wild yet. But you will see them all at the zoölogical garden in Batavia'....

We left Djokja [Yogyakarta] at sunrise, with enthusiasm somewhat dampened from former anticipations of that twenty-five-mile drive to Boro Boedor [Borobudur].... We had expected to realize a little of the pleasure of travel during the barely ended posting days on this garden island, net-worked over with smooth park drives all shaded with tamarind-, kanari-, teak-, and waringen-trees, and it proved a half-day of the greatest interest and enjoyment.... The way was fenced and hedged and finished, to each blade of grass, like some aristocratic suburb of a great capital, an endless park, or continuous estate, where fancy farming and landscape-gardening had gone their most extravagant lengths. There was not a neglected acre on either side for all the twenty-five miles; every field was cultivated like a tulip-bed; every plant was as green and perfect as if entered in a horticultural show. Streams, ravines, and ditches were solidly bridged, each with its white cement parapet and smooth concrete flooring, and each numbered and marked with Dutch preciseness; and along every bit of the road were posted the names of the kampongs and estates charged to maintain the highway in its perfect condition. Telegraph- and telephone-wires were strung on the rigid arms of cotton-trees, and giant creepers wove solid fences as they were trained from tree-trunk to tree-trunk—the tropics tamed, combed, and curbed, hitched to the cart of commerce and made man's abject servant. . . .

With five hundred Buddhas in near neighborhood, one might expect a little of the atmosphere of Nirvana, and the looking at so many repetitions of one object might well produce the hypnotic stage akin to it. The cool, shady passagrahan at Boro Boedor affords as much of earthly quiet and absolute calm, as entire a retreat from the outer, modern world, as one could ever expect to find now in any land of the lotus. This government rest-house is maintained by the resident of Kedu, and every accommodation is provided for the pilgrim, at a fixed charge of six florins the day. The keeper of the passagrahan was a slow-spoken, lethargic, meditative old Hollander, with whom it was always afternoon. One half expected him to change from battek pajamas to yellow draperies, climb up on some vacant lotus pedestal, and, posing his fingers, drop away into eternal meditation, like his stony neighbors. Tropic life and isolation had reduced him to that mental stagnation, torpor, or depression so common with single Europeans in far Asia, isolated from all social friction, active, human interests, and natural sympathies, and so far out of touch with the living, moving world of the nineteenth century. Life goes on in placidity, endless quiet, and routine at Boro Boedor. Visitors come rarely; they most often stop only for riz tavel [rijsttafel], and drive on; and not a half-dozen American names appear in the visitors' book, the first entry in which is dated 1869.

I remember the first still, long lotus afternoon in the passagrahan's portico, when my companions napped, and not a sound broke the stillness save the slow, occasional rustle of palm-branches and the whistle of birds. In that damp, heated silence, where even the mental effort of recalling the attitude of Buddha elsewhere threw one into a bath of perspiration, there was exertion enough in tracing the courses and projections of the terraced temple with the eye.

Even this easy rocking-chair study of the blackened ruins, empty niches, broken statues, and shattered and crumbling terraces, worked a spell. The dread genii by the doorway and the grotesque animals along the path seemed living monsters, the meditating statues even seemed to breathe, until some 'chuck-chucking' lizard ran over them and dispelled the half-dream.

In those hazy, hypnotic hours of the long afternoon one could best believe the tradition that the temple rose in a night at miraculous bidding, and was not built by human hands; that it was built by the son of the Prince of Boro Boedor, as a condition to his receiving the daughter of the Prince of Mendoet for a wife. . . .

Nothing from the outer world disturbed the peace of our Nirvana. . . . Life held every tropic charm, and Boro Boedor constituted an ideal world entirely our own. The sculptured galleries drew us to them at the beginning and end of every stroll, and demanded always another and another look. A thousand Mona Lisas smiled upon us with impassive, mysterious, inscrutable smiles, as they have smiled during all these twelve centuries, and often the realization, the atmosphere of antiquity was overpowering in sensation and weird effect.

Boro Boedor is most mysterious and impressive in the gray of dawn, in the unearthly light and stillness of that eerie hour. Sunrise touches the old walls and statues to something of life; and sunset, when all the palms are silhouetted against skies of tenderest rose, and the warm light flushes the hoary gray pile, is the time when the green valley of Eden about the temple adds all of charm and poetic suggestion. Pitch-darkness so quickly follows the tropic sunset that when we left the upper platform of the temple in the last rose-light, we found the lamps lighted, and huge moths and beetles flying in and about the passagrahan's portico. . . . There was infinite mystery and witchery in the darkness and sounds of the tropic night—sudden calls of birds, and always the stiff rustling, rustling of the cocoa-palms, and the softer sounds of other trees, the shadows of which made inky blackness about the passagrahan; while out over the temple the open sky, full of huge, yellow, steadily glowing stars, shed radiance sufficient for one to distinguish the mass and lines of the great pyramid. Villagers came silently from out the darkness, stood motionless beside the grim stone images, and advanced slowly into the circle of light before the portico. They knelt with many homages, and laid out the cakes of palm-sugar, the baskets and sarongs, we had bought at their toy village. Others brought frangipani blossoms that they heaped in mounds at our feet. They sat on their heels, and with muttered whispers watched us as we dined and went about our affairs on the raised platform of the portico, presenting to them a living drama of foreign life on that regularly built stage without footlights....

At Solo, second city of the island in size, one truly reaches the heart of native Java—the Java of the Javanese—more nearly than elsewhere; but Islam's old empire is there narrowed down to a kraton, or palace inclosure, a mile square, where the present susunhan, or object of adoration, lives as a restrained pensioner of the Dutch government, the mere shadow of those splendid potentates, his ancestors....

The present susunhan of Solo is not the son of the last emperor, but a collateral descendant of the old emperors, who claims descent from both Mohammedan and Hindu rulers, the monkey flag of Arjuna and the double-bladed sword of the Arab conquerors alike his heirlooms and insignia. His portraits show a gentle, refined face of the best Javanese type, and he wears a European military coat, with the native sarong and Arab fez, a court sword at the front of his belt, and a Solo kris at the back. Despite his trappings and his sovereign title, he is as much a puppet and a prisoner as any of the lesser princes, sultans, and regents whom the Dutch, having deposed and pensioned, allow to masquerade in sham authority. He maintains all the state and splendor of the old imperialism within his kraton, which is confronted and overlooked by a Dutch fort, whose guns, always trained upon the kraton, could sweep and level the whole imperial establishment at a moment's notice. The susunhan may have ten thousand people living within his kraton walls; he may have nine hundred and ninety-nine wives and one hundred and fifty carriages, as reported; but he may not drive beyond his own gates without informing the Dutch resident where he is going or has been, with his guard of honor of Dutch

In the Principalities remains 'all the state and splendor of the old imperialism'. (From J. Koning and D. Wouters, *Mooi en Nijver Insulinde*, Groningen: P. Noordhoff, *c.*1927)

soldiers, and he has hardly the liberty of a tourist with a toelatings-kaart....

The young susunhan maintains his empty honors with great dignity and serenity, observing all the European forms and etiquette at his entertainments, and delighting Solo's august society with frequent court balls and fêtes. Town gossip dilates on his marble-floored ball-room, the fantastic devices in electric lights employed in illuminating the palace and its maze of gardens on such occasions, and on the blaze of heirloom jewels worn by the imperial ladies and princesses at such functions.... The susunhan is always accompanied on his walks in the palace grounds, and on drives abroad, by a bearer with a gold pajong, or state umbrella, spreading from a jeweled golden staff. The array of pajongs carried behind the members of his family and court officials present all the colors of the rainbow, and all the variegations a fancy umbrella is capable of showing—each striped, banded, bordered, and vandyked in a different way, that would puzzle the brain of any but a Solo courtier, to whom they speak as plainly as a door-plate....

Quite unexpectedly, we saw the princely personage himself receive his early cup of coffee—attracted first to the ceremony by noticing a man carrying a gold salver and cup, and followed by an umbrella-bearer and two other attendants, enter an angle of the court in whose shady arcade we were for the moment resting. Suddenly all four men dropped to their heels in the dodok, and, crouching, sidled and hopped along for a hundred feet to the steps of a pavilion. The cup-bearer insinuated himself up those four steps, still squatting on his heels, and at the same time balancing his burden on his two extended hands, and proffered the gold salver to a shadowy figure half reclining in a long chair. We stood motionless, unseen in our dark arcade, and watched this precious bit of court comedy through, and saw the cup-bearer retire backward down the steps, across the court, to the spot where he might rise from his ignoble attitude and walk like a human being again.

Eliza Ruhamah Scidmore, *Java: The Garden of the East*, New York: Century Company, 1897, pp. 22–4, 30–3, 38–9, 41–2, 50–2, 58–9, 66–7, 70, 126–7, 129, 132–8, 142–3, 148–9, 151–2, 154–5, 163–5, 174–6, 203–4, 211–13, 241–2, 245–7, 251–2.

17
H. M. Tomlinson Reluctantly Disembarks in Java, 1923

Henry Major Tomlinson, a prolific English journalist and literary man (and author of The Sea and the Jungle*) had thought to avoid Java during his 'journey to the beaches of the Moluccas and the forest of Malaya' in 1923. Gaudy Dutch travel posters had given him the impression that Java was a tourist trap. But when his coasting steamer put in at Batavia, he ventured forth anyway. In Garut, he escaped the heat. And in Yogyakarta, he confronted the magnetic spell of Borobudur.*

THERE was Java. Under some mountains so distant and of so delicate a color that they were only a deeper stain of the sky were spread the flat buildings of Tanjong Priok. To my surprise, I discovered the *Savoe* had a crowd of passengers for Java.... Java was a serious disappointment to me some time before I reached it. The gorgeous East obviously could not be, and ought not to be, so gorgeous as Java's holiday posters, which are in the style of the loudest Swiss art. 'Come to Java!' Well, perhaps not. Not while the East Indies are so spacious and have so many other islands; not if Java is like its posters. Why should the East call itself mysterious when it advertises itself with the particularity of a Special Motor Supplement? I felt I ought to keep away from Java. I thought I would prefer to leave it to those who enjoy traveling round the earth in eighty days, and who see all the wonders of it from the deck of a twin-screw composite restaurant and tennis court.... But Java stood in the way of my coasting steamer, which had to call at every port, and at some places which are not ports but merely wish they were, along the north coast of it. This hindrance had to be endured for the privilege of seeing the outer islands, those unimportant beaches which will have no posters of their own, for which we should thank God and fevers, for some time to come.

Our first Javan port, Tanjong Priok, is the harbor for Batavia. And here the mosquitoes came aboard in hosts so ravenous that they tried to bite their way into the cabins, and so stuck to the new paint.... Java may be a perfectly healthy island, and malaria there—as one is led to infer—as rare and inconsequential as falling upstairs; but the ship's new paint scared me. It broke down my resolution not to see Java, and I fled ashore....

The moist heat of Java's plains and seaports, even when the interest of a place is just a little more remarkable than the temperature, soon turns one to thoughts of escape, in the bare hope that Java somewhere in its garden has a bower which has not the peculiar virtue of a vapor bath. 'Why, if you go to Garut', I was told in a voice which suggested a wonder that I was not required to believe without Thomas's proof, 'you will want a blanket at night!'

'Our first Javan port, Tanjong Priok, is the harbor for Batavia.' (From J. Koning and D. Wouters, *Mooi en Nijver Insulinde*, Groningen: P. Noordhoff, *c.*1927)

I had never heard of Garut, but one place is as good as another to a traveler who is rewarded by whatever he can get. I found Garut in the mountains of central Java, somewhere behind Tjilatjap [Cilacap]. There was no trouble in finding it. Everybody seemed to know it. But I shall remember Garut as I remember Sfax, Taormina, Chartres, Tlemçen, and other odd corners of the earth, some without even a name on the map, where we arrived by chance and disconsolate, and from which we departed with something in our memory, forgotten till then, that had been lighted briefly by what may have been a ray of moonshine. Can such an experience be communicated? But how shall a man define his faith?

Yet there Garut is—or there it was, for I am not going to assert the existence of any spot on earth that was, for all I know, revealed to me briefly by lunar means—there Garut is for me at least, high on a ridge above a confusion of tracks through rice and tobacco plantations. The bearings of the place as I saw it can be only vague, and are probably wrong. The women of its *campongs* cast down their eyes as you approach, the children run into their huts, and the men raise their big hats of grass politely. It is secluded within ranges of dark peaks, but its own fields are bright and have the warm smell of new earth. From a grove of bamboos which fringes the highest ledge of a vast amphitheater terraced with steps of rice you look out into space and down to an inclosed plain. The plain is remote enough to be the ceremonious setting for the drama of a greater race of beings; men would be insignificant there. But the stage is empty; only the cicadas and frogs fill that immense arena with their songs just before the day goes. The ridge of the opposite side of the theater dissolves in rainstorms and is reformed momentarily by lightning.

The night was cold. I had to get up to find that blanket. It was still colder before dawn, when a knocking at my door warned me that I had arranged to go to a village called Jiporai [Cikurai], where it was said I could find a horse and so could ride to some mysterious height where there were hot springs in a forest. That folly was of my doing—it came of my easy compliance to a foolish suggestion made when my mood, induced by Garut and wine, was inclusive and grateful. Nor would it be any use, I found, to ignore that knocking. The native outside was going to keep it up. In the bleak small hours my need for a horse had vanished. I did not feel it. I did not want a horse.

Yet there it was; and out of the village of Jiporai the track mounted quickly, while from the saddle of a stout pony—he was as petulant as myself over this preposterously early excursion—I looked at the darkness which filled the amphitheater of mountains; it was a starless lake of night. We were only just above the level of that expanse of chilling shadow, and its depth, straight down from where the pony's feet clattered the stones on the edge of it, was unseen and unknown. But I stopped him when the sun came. He turned his head, too, with his ears cocked eastward. We both watched. The great space below us quickly filled with light. We could see to the bottom of it. Without a movement, rider and horse were at once placed by the dawn at a dizzy height, on an aerial path.

The pony snorted and shook his head. It was morning, it was warm, and suddenly the earth began to exhale odors. We went through a flimsy *campong*, with relics of totems over its huts, and shy women with babies straddled on their hips pretended they could not see us. The men were in the padi fields beyond. We came upon some of these workers making coffee and cooking rice under a thatch propped on four poles. They accepted us as though they had always known us, and I may have had a better breakfast than that at some time or other, but I don't remember it.

We had to pass through a deep ravine remarkable with the fronds of immense ferns arched from its rocks. There was an outlook at times over steep places, where distant lower Java was framed in immediate tree-ferns, grotesque leaves, and orchids. We passed up to a grassy plateau, which might have been an English ducal park. The flowers there were of another climate—raspberry, brambles, the yellow heads of composite plants, and labiate herbs. In a hollow of a forest beyond there were forbidding and incrusted recesses where the foliage was veiled in bursts of steam. But we did not pause by those caldrons of boiling mud. The smell of sulphur is not good. Those pools were not designed for the rustic wonder of travelers, nor even to admonish them of what follows after sin. I do not know for what they were designed, but a few years ago, so I was told, they boiled over and obliterated forty of the villages where the people are so good-natured.

So we learn that the rich and beautiful island of Java is not, after all, a creation especially intended to support the flamboyant posters of Batavia. Sometimes it does things on its own account. It was even with a degree of pleasure, later, that I learned I should not be allowed to leave Sourabaya without a visit to the port's medical officer. It broke the spell of the Garden of Eden to find that that corner of it is infected by the plague, and that anyone emerging from the garden is suspect of subtle evil. But what attraction would there be in a snakeless Paradise?

When a scholarly English traveler I met at Batavia found in me no warm regard for the ancient Buddhist tope at Borobudor, and only a loose wish to see it before I died, he adjusted his monocle to get me into a sharper focus. I tried to meet his sudden critical interest with an aspect hastily mustered of original intelligence, but I could see my reputation had perished. He was disappointed. An unfortunate sign now showed in me that he had hitherto overlooked. I should have to make a fight for this, or it would end in my seeing the famous curiosity.

The archæologist was severe, and spoke slowly, like one who knows the truth. 'Remember, you have come many thousands of miles. It is probably the only chance in your life to see one of the most remarkable temples in the world.'

'But I don't want to distort my brief span by trying to cram every experience into it. I know I can be happy without Borobudor.'

'Eh? Pardon me. But this relic is unique.'

'So am I, sir; so are you. We are even older than that old tope by about seven hundred years. Besides, I have seen its portraits, and they fill me with despair.'

His monocle dropped from his raised eyebrow. 'Despair? It's a wonderful mass. It's an amazing pile. I do not understand so incurious a man. You astonish me. Everybody sees it who comes to Java. Despair? Why, dammit, think of the industry, think of the devotion of those people who converted the rocks of a hilltop into a huge temple!'

'I have thought of it. That's what disheartens me. It is horrible. Those stones with their carvings commemorate calm and settled error. . . .'

The archæologist held up his right hand and waved it gently in the way a policeman stops the traffic. I stopped.

'Excuse me. If they erred, what is that to us? It is their art which matters, not its cause. And even its cause—such devotion to their faith, carried to its topmost pinnacle, crowning a hill with the proof of man's yearning mind, deserves more than your flippant indifference. Borobudor was a temple to God.'

'That's the trouble with it. That's the nearest to God we ever get. If I went, I know I should only have it rubbed in that we seem unable to receive a simple lesson in wisdom without at once beginning to elaborate it into a system to justify what we want to do. Why, if Buddha were to come to Java, Borobudor is probably the one place he would be careful to avoid!'

'Sir, you have no reverence for your fellows. Those sincere stones celebrate faith.'

'What of that? Have you ever had to jump for it when faith dropped one of its sincere bombs? Those fervid stones embarrass me like fetiches and patriotic songs.'

'I've never heard anything like it. You can't mean what you say. I won't listen to you. Here is man struggling upward through the ages, and he reaches something so wonderful as Borobudor, and you talk as though it were a symptom of a disease. Monstrous! Why ever do you travel, sir?'

'I don't know, but I prefer to enjoy it, if I can.'

'But how can you enjoy it if you miss the most important things in travel? What do you learn?'

'The confirmation of my prejudices, I suppose. What else could I learn?'

'Ah, the war has destroyed respect and worship in you younger men. You want to begin everything again, but you cannot, you cannot. Borobudor is there, and it is too strong for you. You think you can forget it, but it won't let you.'

'Why was that temple ever dug out of the jungle to which it belonged? A fixed tangle of dark and incurable thoughts!'

'I cannot discuss this with you, sir. There is nothing else like Borobudor in the world.'

'This poor world is overloaded with Borobudors. We struggle beneath them, yet nothing will satisfy you but you must dig out another from the forest which had fortunately hidden it. There it is, to distress any wretched traveler who passes it with the idea that man will never escape from himself. Borobudor is a nightmare.'

Shortly after this interview I was on my way to the ruins. It was not by intention. My affairs drifted that way, somehow, perhaps because the roads of the world are clandestine with their memory of the past, and so we move in the old direction of humanity without ever knowing why. The archæologist was right. The Borobudors are too strong for us. There is no escape....

In the hotel at night I argued myself into a corner. What about visiting this temple, and doing what all travelers do? At length I surrendered. It was clear I could not dispute with everybody who asked me what I thought of Borobudor—for it would come to that; it would be assumed that I had seen it

'The Borobudurs are too strong for us. There is no escape.' (From Eliza Ruhamah Scidmore, *Java: The Garden of the East*, New York: Century Company, 1897)

and was awed—so I ordered a conveyance for the morning. The temple was thirty miles distant. I had better get it over.

In the morning, before my very door, the folk of Jokyacarta [Yogyakarta] were going to market. They were gossiping, and looking this way and that, as casual as though all days were holidays and obsessions of the human mind had never been perpetuated in monumental stone and enduring empires. Nothing was dark in that throng. It was as varied as a garden and as engaging as birds on a June morning. I joined it. It is pleasant to go to market, for markets are places where people live and where even small change is more important than lost and awful fortunes. A motor-car stood at my door and its driver salaamed, but I gave him good morning and passed on.

Jokyacarta, even to my inexperienced eye, was an important city, for its people seemed unaware of the urgency of the outer world. They were going their own way. . . .

How faithful and bounteous a spring of original life—common humanity everywhere! It is like rain and grass and the sun. To read the history of Java from the Hindu to the Dutch would lead a distant student to imagine that its people must have had every spark and airy bubble compressed out by the strong governments of fifteen centuries, and that there is left but a flat and doleful residue of homogeneous obedience. But some joy will remain after the strongest governance has done its best. That market place, though at least as old as sultanic prejudice in concubines, and vastly more ancient than the Prophet's victories in Java, might only that morning have come into bloom. Its leisurely throng had forgotten that the military roads of their island garden are built of their bones. They have risen again. . . .

That was an assurance worth going to market to get. It was procurable at any stall in Jokyacarta. What you bought was wrapped up in it. It was certain these people could do very well without the aid of sultans, priests, or governors, whose assistance, somehow, they had survived, so far. What promise of Java's development could be better than the figures and poise of those girls who were selling their *batik* to a Chinaman? They were there when the Hindus came, they had survived every conquest, and there they were still, with eternity in which to have their own way if Chinamen were obdurate. Their gestures and pose proved them to be of an ancient and noble line. The great dove-colored bull beside them, his nose and ears of black velvet and his eyes tranquil with drowsy pride, had come with them all the way from the past. That market place, with its craftsmen and women, could light the fire of humanity again on the abandoned hearths of a bare continent.

I was already settled with this comfortable thought in the train for Sourabaya next day when an official from my hotel, whose anxious face was peering into every coach, presently found me. There had been a mistake. I had not paid for my motor-car to Borobudor. It had waited for me all day. Borobudor?

Henry Major Tomlinson, *Tide Marks: Being Some Records of a Journey to the Beaches of the Moluccas and the Forest of Malaya in 1923*, New York: Harper and Brothers, 1924, pp. 88–90, 104–14.

18
Augusta de Wit Romanticizes 'A Native of Java', 1923

Born and raised in the Dutch East Indies, journalist and author Augusta de Wit wrote knowingly of Java, although often in a highly romanticized fashion. In the passage below, for example, she describes how a 'brown man of the earth' labours not 'for wages in the service of the alien but for his own ends, at his own will, in his own way'. Unlike some of our travellers, however, de Wit was aware of colonialism's harsh side. In the story from which this passage is excerpted, the 'native of Java' in question is actually a convict condemned to deportation and, it transpires, an early death.

AS manifold as the loveliness with which it caresses his senses and his soul, so manifold is the love with which the native of Java, the brown man of the earth, living poorly and humbly close to the glebe, loves his land, most lovely Java.

He loves the soil of Java, which is a fire in the east monsoon season, and a flood in the months of the rains; which, under the tread and turning wheels of the long files of bullock-carts, slowly creaking along the harvested sugar-cane fields, floats up in clouds of whitish dust, and lies dark and cool on the hands of the women at work in the rice-field, transplanting the month-old seedlings from the seed-plot to the sawah, as they carefully press the soaked earth around the limp, pale green stalk that the plant may strike firm root and thrive; the earth that softly yields to the potter's fashioning fingers as he shapes the lump on his revolving disk into capacious rice-bowl or slender cooling-jar; the earth that stands steadfast and strong in the dikes of the flooded terraced fields, bearing up against the hillside a flight of lakelets, crystalline pools, where the purple skies of sunrise and sunset and all the sailing clouds of azure noon float reflected amidst the green of the sprouting rice.

He loves the scents of Java, the thousand scents that float on the passing breeze; the smell of wet earth and boulders in the shallow river, of young leafage springing from shoots all

'He loves the soil of Java ... the earth that stands steadfast and strong in the dikes of the flooded terraced fields. . . .' (From J. Koning and D. Wouters, *Mooi en Nijver Insulinde*, Groningen: P. Noordhoff, *c.*1927)

swollen and gleaming with sap, of the pasture where the naked herdsman lies piping in the shade as his broad buffaloes plunge into the pool, snorting; the bitter smell of the tall alang-alang grass afire on the hillside, where some reckless nomad sits waiting to sow his rice in the ashes, that he may gather a harvest from soil unbroken and untilled; the exquisite fragrance of the penang [areca-nut] palm in bloom, breathing from the wall of trees that hides from view a hamlet; the pungent smell of the market-place and the crowded highway, and the home where children are always being fed; the odour of the incense that hallows the eve before the Day of Prayer; the scent of the white jessamine wreaths that crown the bride and the bridegroom sitting in state at the wedding feast.

He loves the sounds of Java, the innumerable sounds, to which his heart makes gladsome answer; the delectable sound of the rain on the living leaves of the wood and the withered leaves of woven roofs, on the boulders in the ravine, where the silent brooklet begins to purl and cluck, and suddenly lifts up clear voices calling aloud; the wind-stirred rustle and murmur of the bamboo grove that surrounds the village, a swaying cloud-like wall of foliage, where at sunset swarms of rice-birds twitter unseen; the busy sound of rice-pounding, a dancing rhythmical beat from the hollowed-out tree-trunk lying between the starrily flowering citron bushes of the fruit garden, the scented space of shade and freshness for the labouring housewife, over whose shoulders the babe, cradled in the deftly slung slendang cloth [shawl], laughs at the dance of the golden rice grains bouncing away from the pestle; the festive sound of the gamelan orchestra played by an able musician whose touch on the bronze sends the deep tones and the shrill soaring through the silences of the night, and they hover for a while firefly-like, gladdening hearts far away.

He loves the colours of Java, the clear and effulgent ones, the darkly glowing ones saturated with mellow sunshine, the delicate ones, tender and cool as moonlight upon dew; the tints of diaphanous hilltops in the distance and of the mist-flushed plain; the sparkling green of the sawah [paddy fields]; the purple in the heavily hanging blossom cone of the banana-tree at the back of his house; the thousand-tinted sparkle with which the long files of gaily attired women strew tamarind-shaded roads that lead to the market; the scarlet and blue and green and black and gold of the garb in which gods, nymphs, and heroes are represented in the solemn wayang drama.

And he loves the daily labour of Java, the labour he does not do for wages in the service of the alien, but for his own ends, at his own will, in his own way, the ancient way of his people, as his father taught it to him.

To go to the sawah at daybreak—the field that out of the broad expanse of common land the headman and elders of his village have allotted him as his own for a year's space—to go, his feet in the dew-frosted grass where scents still are asleep, his face lifted to the colourless sheen that precedes dawn, to move lightly through air fresh and abundant as welling water, feeling on his shoulder the light burden of the wooden plough, and looking at the yoke of broad-backed buffaloes that slowly tread the wonted way—the ploughman's good friends they, who lend their strength to his knowledge;

to drive the long furrow through soil growing warmer as morning glows into noon, the rich soil where the rice grains of the recently gathered harvest already are sprouting under the ashes of the stubble fires (the ploughman thinks of the frolicking boys of the village, how they leapt among the leaping flames, and remembers himself leaping and frolicking thus not so many harvests ago); to return in the heat of noon to find the coolness of the house, and the meal of rice and dried fish neatly served on a strip of freshly gathered banana leaf; to see, in the slowly cooling hours of the afternoon, the lengthening shadows gliding along a narrowing strip of unbroken ground, slender man's shadow side by side with broad shadows of plough beasts; and to draw the plough out of the last furrow as he sees the buffaloes and his own arms and knees reddened by the glow of sunset. Thus to live through the working day is sweet to him....

He loves the feasts of Java, at which gods and genii are his unseen guests. For many weeks beforehand he rejoices in the coming harvest feast. A merry sight it is to him when the housewife prepares and dyes with yellow boreh powder the sacrificial rice and chooses the finest fruit of the garden for an offering to Dewi Sri, the Rice-Goddess. When the angkloong players begin the feast, shaking their sets of graduated bamboo tubes from which the liquid notes pour forth clucking, he binds a handful of rice-ears into his kerchief, and another handful into the kerchief his friend holds out toward him, and, with the gaudy bundles dangling from either end of his bamboo yoke, he performs the graceful dance that follows the rise and fall of that undulating music.

He loves the graves of Java, the miracle-working tombs of the saints of Islam and of mighty sultans, whither he goes to pray for a blessing on his undertakings; the graves of his father and mother, whither on set days the household brings flowers and food, that the souls of the dead may feed, and rejoice, beholding their children's love still faithful to them and mindful of their needs in the cold Land of Shadows.

All things of Java he loves; all of his lovely country is sweet to his senses and sweet to his soul....

Augusta de Wit, *Island-India*, New Haven: Yale University Press, 1923, pp. 29–32.

19
Helen Churchill Candee Observes the 'Conquering Race Eating His Mid-day Snack', 1927

American writer and traveller Helen Churchill Candee introduces us to the delights of the Java hotel in the 1920s and to the ways and means of Indies bathing, sleeping, and eating. Surabaya, she tells us, 'is hot'.

IT is small wonder that a place is remembered by its hotel, for a hotel is the first experience in a foreign land; it is impressed upon the mind which is fresh and purposely open for impressions. And this is what a hotel in Java is like: a smallish building of two stories high, flanked with long rows of tiny cottages built in solid blocks. The central building uses its height to make lofty its rooms, usually a lounge, a writing-room, a dining-room. The cottages outside are the bedrooms. They look like the workmen's bungalows run up at little expense in the neighbourhood of some big industry.

But enter one and realise its delights. It is a bedroom. You ask for—you pay for—a bedroom and you get a cottage. Marvellous. The entrance is a porch sitting-room fitted with a centre-table and lamp, a desk, a cane chaise-longue, and a rolled curtain which conceals or reveals at one's pleasure. Back of this charming place is a shady bedroom, with net-draped bed. And back of that is a commodious bath-room of the tropic plan.

Outside are palms, the trunk of a tree filled with orchids, vines of jasmine and bougainvillea. It is a hot country, you see, and the hours from eleven to five are spent in this bedroom-cottage where the frequent bath and much sleep fill the stewing hours.

The bath is amusing. It is staggering at first in its novelty. A tub is no part of it. The whole room is the tub. In one corner is built a reservoir, or cistern, the tipped floor of cement contains a drain. The reservoir is full of water and beside it stands a generous dipper. All these things convey that one is to stand and sluice with cold water, there never being a hot water tap. And one becomes charmed with the refreshment, and the allurement increases, until one falls under the spell and is drawn bathward several times a day.

Even the bed cannot keep the occupant from the bath, the bed which has its own strange and suitable peculiarities. Its mattress is thin and hard but acceptably flat, without the impress of a previous occupant. The scant pillows are featherless but stuffed with the cooler kapok cotton. Then, laid lengthwise in the middle of the bed is the short frilled bolster, jocularly named a Dutch wife.

At five or so in the afternoon a soft-footed 'boy' comes in with a long brush of switches with which he sweeps the mattress to its most secret corners, and then closes the net and tucks it in exceeding tight. Nice boy, to be so tidy as to brush the bed. It is when the sleeper puts himself to bed a few hours later that the discovery is made that there are no bed-covers—not even a top sheet! But that is the way in Java, the country being hot by night as well as day....

In the dining-room a sort of play goes on at tiffin time.

A very large and self-important Dutchman is seated alone at his table. Two waiters come in, each bearing two silver dishes. They bend and offer him of their delectables. He heaps a huge soup-plate with dry and snowy hot rice. On a side-plate he lays a fish, a slice of ham, a fritter. Two more boys side by side approach him, each bearing on right hand and on left a silver dish, from all of which he serves himself.

Another turbaned boy approaches, and another, and yet another, until there stands a waiting procession of boys with viands. In all there are twelve—and that makes twenty-four *plats.* And one man takes from them all! It is a feat, an unbelievable preposterous joke. Not at all—it is one of the Conquering Race eating his mid-day snack. It is the Dutchman's Rijsttafel, which is translated, rice-table.

And when the gentleman has been swelling at this varied meal for twenty minutes, the entire procession appears again and the eater again heaps twenty-four varieties upon his depleted plates. It is not believable, yet it is true, and may be seen any day in any restaurant of Java....

Sourabaya is hot. That is not denied even by those who are making fortunes there. It has seated itself on the flats by the Sunda Sea and the sun takes mean advantage of its necessity which is sea commerce. It is in a way a rival of Batavia as a shipping town. Preposterous figures are given about tonnage in and out, which seem to mean far less than that it is a port of call for all the Australian and New Zealand ships.

At the big docks—which are according to rule at the far end of miles of swamp—lie vessels on which are painted names of astoundingly unreal places like Tasmania. They look like trans-Atlantic ships with tiers of cabins amidships and a general air of dominating the seas. For some reason they never seem to be mentioned in the banks, hotels, and offices. But the name KPM is fluttering like a national flag.

Sourabaya is as busy a place as though its air were crisp and frosty. Commerce and shipping. A fever of selling pervades the place. Banks, insurance companies, shipping firms, fight fiercely for supremacy, and men work at top speed while down in each one's heart the real desire is for a chaise-longue in a cave and a long cool drink.

Up the street where the hotels are, the work still goes on. Hotel rooms are dark caves, but each has an open-air sitting-room, and these in a row make a lesser business street. Utter frankness prevails. A man in pyjamas dictates to one in an undershirt and trousers. A man on a chaise-longue reading legal-cap, throws open his dressing gown above and below the belt to prove his innocence of undergarments. Another in shorts forgets his stockings so busy is he with his typing. All the lush crops that grow on this lush isle are bought and sold in this little alley of bed-rooms fed by a tiled sidewalk two feet wide in the breathless court of the hotel.

Motor agencies everywhere sell cars in beguiling terms, but carriages are not extinct. There is one called deliciously Mylord. At its name you fancy a powdered footman bowing his master in. But the reality is a shabby carriage drawn by sad small horses. The real delight is a cart of one tiny horse and speechless native driver, a dos-à-dos called sado. It is cushioned, it is canopied, and moreover it is decorated with bright silk sashes which float out behind with hilarious gaiety. One mounts at the back, and instantly the thing tips backward to its balance, the whip cracks, the pony slowly re-collects his legs and the vehicle joins the throng, while from the sober passenger's back float the pink and yellow sashes. The man is paid (forty cents an hour!) out on the street, for no hotel would

'The real delight is a cart of one tiny horse and speechless native driver. . . .' (From M. T. H. Perelaer, *Het Kamerlid van Berkenstein in Nederlandsch-Indie*, Leiden, 1888–9)

allow vehicles of such picturesque delight to round the curve to their carriage block where motors pant and growl.

This trap takes you to post and telegraph if you give the word Kantoor, the one Dutch word they know and in driving thither the way is down the street of pianos, motor companies, European clothing to a crazy Chinese quarter full of delicious colour and care-free people. . . .

Ships at Sourabaya go everywhere. Big vessels of luxurious travellers are painted white or pearly grey and bear an Australian name, or a city of New Zealand. But deeper romance hovers around the dark and rusty hulls of the sea-rovers whose names are written in Chinese characters as well as in Roman. They hail from Borneo, Celebes, Thursday Island, and are going to Amoy or Polynesia—or to the Americas. Never is one journey like unto another and each is full of adventures in those parts of the earth and the sea where the Tables of Stone fail to extend their influence.

It is at Sourabaya one can take ship for Singapore—or straight up through the islands to Hong-Kong.

Helen Churchill Candee, *New Journeys in Old Asia*, New York: Frederick A. Stokes, 1927, pp. 191–4, 230–2, 234.

20
Charlotte Stryker and Family Weather the Monsoon, 1920s

As a teenaged girl, Charlotte Stryker accompanied her parents and four younger brothers to Java, where her father tried unsuccessfully to establish a tapioca (cassava) plantation. They arrived from Roselawn, Pennsylvania, armed with thirty-one pieces of hand luggage, a car, and 'seven immense crates of canned goods, five large trunks of clothing, a first-aid kit complete with anti-snake serum, and a dozen bottles of cod liver oil'.[1] *Their little community somewhere along Java's west coast also included the overseer and assistant overseer, Mijnheers Klaut and Vos, engineer John Westbrook and his Vassar-educated wife Helen, and Miss Blue, an English governess.*

MIJNHEER KLAUT had been quite right about the weather. The two hours' rain of early fall had lengthened to an afternoon's downpour. Then, often it would rain all night and sometimes all day as well. When the sun shone it was as bright as ever but did not dry the ground which became dense, unyielding mud.

The great trees still were felled but could not be burned or moved. Entangled with ferns, vines, and undergrowth, they lay in a morass flecked in low-lying places with puddles and pools. Dad struggled like a Titan against the rain and the mud. But by December only fifteen hundred of his ten thousand acres had been cleared.

The rain was not cooling. The temperature remained above ninety—day in, day out, and night and day. We lived in a steaming miasma of heat that grew fungus in our shoes during the night. Our clothing broke out in spots and rotted before our very eyes. Each time the sun appeared, everything was rushed outside to dry but nothing ever did dry completely.

As though seeking refuge, armies of insects invaded the house. We put small cans of kerosene under the legs of

[1]Charlotte Stryker, *Time for Tapioca*, New York: Thomas Y. Crowell, 1951, p. 6.

cupboards and bureaus to try to save our clothing from the omnivorous ants and beetles. Our dozen or so geckos were joined by so many friends and relatives that they collided on the ceilings and fell with a thump on top of the mosquito-net frames. . . .

After the failure of our canning enterprise, Mother gave up and let the rain and weather have their way. Idle and easy-going by nature, she was not one to struggle too hard against circumstance.

When our trunks had arrived from Batavia, Mother had been glad to abandon her explorer's costume and dress nicely again. But that had been months ago, before the rains came. Now she was in her sarong and waist all day and spent most of her time reading on the sofa.

Mevrouw Klaut and Mevrouw Accountant Aken dropped in occasionally for morning coffee or for afternoon tea. Mevrouw Assistant Overseer Vos, apart from a formal monthly call, did not appear at all.

Though constantly on hand, Miss Blue wasn't much of a companion. She had at first been extremely ready with her advice and suggestions. Mother had listened mildly but somehow just didn't get around to putting them into effect. So Miss Blue's comments had ceased. Mother just wouldn't take the white man's burden seriously. Obviously she was not of empire-builder stuff.

However, in the oppressive afternoon heat, even Miss Blue was glad to abandon part of her own burden. By mutual consent, we skipped our afternoon classes and all took a good long nap. After that, we went bathing in the river which was one form of activity that was not interfered with by the rain.

Like Mother, we became extremely negligent in our attire. The boys forgot all about shirts and wore only shorts. We went barefoot in the house and kept our shoes on the porch as Mother insisted on our wearing them out of doors through fear of hookworm. I wore only a sarong and looked, as Miss Blue frequently remarked, just like a native. But Miss Blue herself kept up her standard. She never left the house without her cotton gloves and her big hat, its blue plume sadly bedraggled but still bobbing genteelly.

At first Mother had struggled against Cook's tendency to feed us only *rijst tafel* but now she had given up on that point too. We had rice for lunch and dinner and sometimes for breakfast as well. We got used to it and liked it. Canned goods and similar delicacies had disappeared from our menu. Also things we had thought to be necessities such as bread and potatoes, beef and pork, and fresh butter and milk. Our provisions all came from the local market and Mother had forgotten about supplies from home. Her only importations now were cod-liver oil and fresh batches of detective stories. . . .

Helen said, now that the weather was so bad, one must make a special effort not to get into a slump. But despite Vassar, advertising agency, vim and vigor, she slumped in the rain just as everyone else did. And she slumped and lolled around in a sarong, too. At first she said sarongs were sloppy and that going without pants made her feel indecent. But she soon found that being pantless, as well as shoeless and stockingless, was cool.

Helen and I had long since abandoned our mathematical studies as hopeless. As to Latin, Helen said I could get along with it just as well if I studied by myself. Perhaps I could have if I had tried, but detective stories were so much more interesting than the Gallic War. Helen thought so too and spent most of her time on the sofa reading them.

But shortly after New Year's, she was literally forced to get up. Long ago, when the rains began, the pretty chintz slipcover had rotted, spotted, and been thrown away. The blue velvet upholstery had been very hot to sit on until it became thoroughly damp and stayed that way. Then holes appeared in the sofa and in the big armchairs also. Helen thought they were made by the insects that flew around the house or by the invading armies of ants and beetles.

But one evening she found out what caused the holes. A long, pale worm ate his way out of the sofa and right into her lap. The furniture was full of them and they soon ate their way out into the open and stayed there, sticking clammily to the tattered blue velvet. . . .

So Helen got rattan furniture that raised welts on your back. Also, her upholstered bedsteads had to be replaced by iron ones with mosquito net frames.

The dining-room furniture did not have to be disposed of. After John had spent a Sunday with a blowtorch, removing the varnish, it was all right. Then Gee could get the plates off the table without using both hands. And Helen and John could get their elbows off without leaving behind a piece of skin....

By the time the rains began to lessen in the spring, everyone felt more or less like crying. But all things come to an end, even the Java rainy monsoon. It was Mother who revived first and felt, despite her natural indolence, that she must have something to do.

Charlotte Stryker, *Time for Tapioca*, New York: Thomas Y. Crowell, 1951, pp. 119–20, 144–5, 216–18.

21
John C. Van Dyke Admires Malang, 1929

'The Dutch have built modern Malang and the natives have supplied the color,' writes American art critic and scholar John C. Van Dyke. Van Dyke finds much that is picturesque in Java of the late 1920s, but what he finds most impressive is the Dutch colonial administration. Like J. W. B. Money of nearly eighty years before (see Passage 10), he thinks other colonial powers could learn a thing or two, including the United States, whose own South-East Asian colony was near by in the Philippines.

MALANG! Who of Europe ever heard of it? What tourist folder or red-backed guide book ever sounded its praises about the world? And what would one expect of a town in the interior of Java, with a population of forty thousand people, and thirty-six thousand of them natives? I have been in towns of that size in Mexico, South America, Australia, the United States and I have gotten out of them as quickly as possible. There was apparently no good reason for their existence and it would seem as though no good thing could come out of such Nazareths. But Malang, up in the foot-hills of Java, is quite another story.

It is a city in a wide valley surrounded by great mountains, a city of fine streets, with overhanging foliage, parks with huge waringin trees, good houses, lawns, flowers, a picturesque river with high bridges, a fine climate, and lastly (but not least to the tourist) good hotels, good drinking water, and a fine swimming pool....

The Dutch have built modern Malang and the natives have supplied the color. Here as elsewhere in Java, the *sarongs*, the scarfs, and the head gears of the natives light up the streets and markets and make of the whole place a series of pictures. Sitting under the trees in the great park (*aloun-aloun*), drifting along the streets, working in the saw mills and blacksmith shops, bathing in the river, kneeling in the mosques, the natives are always picturesque. Oh! they have their defects, no doubt. If any one wishes to insist that some of them have a way of postponing work from day to day, and that the women have the unhappy habit of chewing areca-nut preparations, no one will deny the statements. But there may be a fling-back to the effect that the chewing-gum habit in America runs into the millions, and that the bread line never grows less. And there never was any saving grace of picturesqueness about either the American shop girl or the American beggar....

Now this town building has not been mere aimless energy touched up with splashes of color. You will find in streets, houses, bridges, parks, shops, an attempt, at least, at adaptation to climate, soil and tropical conditions. The Dutch, naturally enough, brought here from Holland many ideas that were well suited to the Netherlands, but not applicable to Java. They are still getting modern ideas of building from Holland that might better be left in Amsterdam and The Hague. But from the first, even when they were putting up classical palaces in brick and stucco, there was an attempt at adaptation to local heat conditions. Tree-planting, park and avenue planning, water-ways and road-ways in shadow, were from the beginning. And all the cottages and villas have been broadly roofed, deeply shadowed, built widely open to the air. The Netherland ideas have been modified and even transformed into something that is well suited to Java.

The imposition of foreign ideas upon a country or a people is always a questionable proceeding. They are usually not suitable, and often meet with native opposition for no other reason than that they are foreign. The Dutch have learned this from experience and are now disposed to accept local conditions. They are protecting the native and allowing him to develop along his own lines. They are not trying to change him into a Dutchman. Even in the small matter of costume they have not asked him to wear Dutch shoes, shirts and hats, but allowed him to go his own way barefooted and in *sarong*. His religion, customs, traditions are respected; he is protected in his land holdings, helped in his agriculture, and heard in legislative and political proceedings. Largely perhaps by reason of a liberal and generous policy in administration the Dutch have proved themselves the most successful colonial administrators of modern times. Both the United States and England could learn much from them, if they would....

It is hardly worth while hemming over some minor errors in the Dutch administration of Java and the islands. The fact is that the Netherland East Indies are well managed, better managed than any colonies elsewhere on the map. The American administration of insular possessions may approximate that of the Dutch in efficiency, but it is carried on at a financial loss. The Dutch are indirectly making money out of Java, but they are letting the natives make money, too. Moreover, they are putting back into the country millions in development. They are trying to establish a just and equable government and a prosperous colony. To that end they are confirming the land rights of the natives, introducing improved methods of irrigation and husbandry, conserving the forests, establishing native schools and universities, building cities, roads and bridges, opening up new transportation routes, and doing a thousand and one things looking to the betterment of town and country. The result is the natives are well fed, well housed and dressed, look happy, seem contented. And Java is a joy to the traveller, the most delightful of all the tropical countries. The Dutch must receive the credit for much of this. Why not say so without reservation?

22
Harriet W. Ponder: The World's Model Colony is a Motorist's Paradise, 1935

Englishwoman Harriet W. Ponder lived in Java in the 1930s. Her writings represent the epitome of a trend in which the island and its indigenous people were depicted as a huge tableau vivant *to serve as a backdrop for the main actors: the Dutch, of course, and also 'British, Germans, Danes, Swiss, and Americans, with their commercial enterprises that are the very soul and spirit of modern history'.*[1] *Here she describes the modern Indies Dutch for whom Java is home and introduces us to the vexations and pleasures of touring the island by car.*

ALTHOUGH Java and Malaya are next-door neighbours, and both are tropic colonies, the life that Europeans live in each of them is absolutely different.

The most notable difference of all has nothing whatever to do, so far as one can see, with the essential difference in nationality, for nationality does not explain why it is that Malaya, except for babies and tiny tots up to six or seven, is empty of European children, whereas in Java they abound. It is difficult to see why the fact of being born Dutch should render children immune to the tropic sun; or why, on the other hand, the British child should be more susceptible to it.

Both colonies are close to the Equator: on sea-level they have much the same climate; if anything, Singapore is rather

[1]Harriet W. Ponder, *Java Pageant*, London: Seeley Service and Company, 1935, p. 300.

cooler than Batavia. Yet it is the firm conviction of the British in Malaya that children cannot possibly be brought up there; and the equally firm conviction of the Dutch that their youngsters will be every bit as happy and healthy in Java as they would in Europe. And the fine healthy young Hollanders, sunburnt and sturdy and full of life, that you see on all sides as soon as you land in Java certainly give the lie conclusively to the theory that healthy children cannot be reared in the tropics.

In Malaya the sun is regarded as the child's most dangerous enemy. It is only in the early mornings and late afternoons that the Chinese 'amahs' (nurses) are allowed to shepherd their charges out of doors. And even then the poor mites are made ridiculous as well as miserably uncomfortable by their fond parents' insistence on their wearing sun-helmets....

The pity of this sunless treatment is that it turns the children into pasty-faced, peevish, pathetic little creatures, who seem to grow pastier and more peevish every day, until the time comes when (to everybody's relief) their mothers take them off to England, to deposit them with accommodating grandparents, and come back without them.

One sometimes wonders (quite in a whisper!) whether this custom of 'dumping' the children at home was really instituted for their benefit at all? It certainly does leave Mother so delightfully free for her busy life of bridge from ten till one, sleep from two to four, tea-dance, golf or more bridge, cocktails (known here as 'pahits'), dinner, dance—and so to bed; which is roughly the routine of the average Malayan British matron.

Dutch parents, on the other hand, regard the sun as a friend, not as an enemy. Their children run about and ride their bicycles in the sunshine, bareheaded, barearmed, and barelegged, not only in the hills, but in the towns on sea-level, and they thrive on it.

By way of demonstrating their confidence in the East Indies as a place in which to bring up children, the Dutch authorities have transported there the selfsame system of State education that is in operation in Holland. There are schools in every European town in Java, housed in admirably planned modern buildings, with spacious, lofty classrooms open to the air, and glorious gardens and playgrounds shaded by great jungle trees.

The curriculum is precisely the same as that in use in the Netherlands, with the addition of the necessary extra languages....

Dutchwomen are wonderful housewives, as all the world knows, and happily they have brought their talent for homemaking with them to the Far East. I have read (and chuckled over) novels about Java, in which the heroine (obviously 'no better than she should be', but the authors assure us that we could scarcely expect anything else in that 'awful climate') lived a limp, perspiring life, all languor and lovers, perpetually arrayed in a kimono.

I am sorry to disappoint you, but the Dutchwoman in Java isn't in the least like that. On the contrary, like the virtuous lady in the Bible, she looks well to the ways of her household; and if you should happen to catch sight of her in her kimono, she is pretty sure to be either on her way to a bath (we have several a day in Java) after some especially strenuous piece of activity in the kitchen, or else she has donned it for the afternoon siesta, which is the custom here, as in so many other hot countries.

I have never yet come across a Dutch housewife who had fallen into the fatal habit of leaving everything to the 'boys', which is so dangerously easy to acquire where labour is plentiful and fairly efficient. She has one goal in Java as she has in Holland, and that is 'Perfection'. And because modern Dutch homes in the East Indies are very charming, and she has everything to hand to make them more so, it may be that her housepride is, if anything, even greater than it is in her home country.

The new Colonial Dutch style of architecture is designed on an altogether more modest scale than the old. But it is just as good of its kind. The rooms are lofty, the floors tiled, and the windows large; there are usually bathrooms to each bedroom, set-in basins, and all the most shining and up-to-date of electrical devices.

Every bride migrating to the East is quite certain to bring some old family china with her. Brass, Oriental pottery, Chinese embroidery, Batik, and all such decorative adjuncts are plentiful and by no means dear, on the spot; so that the Dutch matron's task of making the 'Home Beautiful' in Java is a happy one.

Every house has its 'voorgalerij', a wide front veranda, which is really a sitting-room open to the air on three sides. This voorgalerij is the pride, the shop window, as it were, of every house, large and small, for, having no front wall, except perhaps a stone railing and pillars, it is in full view from the road. The walls are hung with highly polished brass, and old blue plates and dishes. There are deep armchairs and lounges, covered with soft, inviting cushions; tall palms, and masses of maidenhair fern in pots, and jars of crimson and flame-coloured cannas (for whatever flowers there may or may not be in your garden, the canna is always in evidence all the year round).

At night the houses look even more alluring; and if you walk along the road in the European quarter of any town after sundown you will see every voorgalerij aglow with deep orange or rose-colour or gold, from lights shining softly through native-made silk lampshades.

Gates of any kind are the exception rather than the rule at the entrance to Dutch homes in Java, and the fences or hedges that enclose them are usually quite low. The gardens, like the houses, are their owners' pride. It is easy to make them beautiful, for labour is cheap, and the Javanese native, a descendant of countless generations of cultivators, is a gardener by instinct. . . .

The Dutchman's tropical kit is not quite the same as that of the Englishman. He eschews the collar and tie of our convention, and the jacket of his white drill suit, instead of the usual turn-down collar and lapels, has a stiff, straight band fitting tight round the neck like a bandsman's uniform. It is known as a 'badjoe toetoep' (literally, a 'shut-up jacket'), or 'toetoep', for short, and is fastened round the neck with two brass studs.

That is to say, it is designed to be so fastened. But 'India' (as Java is most often called by the Dutch, to the great puzzlement at first of the English visitor) is a warm country, and Dutch necks are seldom slender. So it is rare indeed to see a toetoep that does not gape widely apart at the neck, with two bright brass studs adorning one side, and two empty studholes on the other.

When a native (by virtue of a generous education system) has attained his heart's desire of 'kredja toelis' (writing work, or, in other words, a clerk's job), he also adopts the toetoep, regarding it as the outward and visible sign of his clerkly attainments. Worn as he wears it, it is both smart and becoming, and I am afraid it must be admitted that his figure is a good deal better adapted to it than those for whom it was originally designed.

In post offices, and all municipal and government offices, these dapper young native clerks abound, cool, immaculate, and slender, in spotless starched white drill toetoep, Batik turban and sarong, and highly polished brown European shoes. Java is not hot to them. Why should it be? And at the end of the long day, which leaves the white man damp and crumpled, the 'djeroe toelis' (clerk) is still as cool and starched, and his collar as neatly fastened, as ever....

There are other contrasts between Java and Malaya beside the reaction of their respective children to the tropic sunshine. The whole 'atmosphere' is different. There is a homeliness and simplicity about Dutch life in Java that is entirely lacking in the British colony over the way. The Dutch love their homes and really 'live' in them, not, as do so many English in the East, regarding them merely as places to sleep in. This homeliness is especially noticeable in the country towns; there is something pleasant and old-fashioned about life in them, despite all their modern comforts, a restfulness sadly rare in these days. The Dutch are 'domesticated', as the English used to be a generation or two ago. You will often see young married couples out for walks in the cool of the evening, with 'Vader' pushing the baby's pram, and perhaps a youngster, the next size larger, toddling at 'Moeder's' side.

The reason is that the Dutch really 'settle' in Java as we do (or used to do) in Australia and South Africa and New Zealand. They are not, like the English in Malaya, perpetually recovering from one 'leave' or getting ready for the next, with always at the back of their minds the thought of retiring to England as soon as ever they have qualified for a pension.

It seems to me a strange ambition. And the problem of why anyone should ever want to leave the eternal summer of Malaya for the almost eternal winter of England remains unanswerable! Half of the unlucky optimists die of pneumonia before they have enjoyed that longed-for retirement for a year, but even that doesn't seem to teach the poor dears common sense! The Dutch have to serve a very much longer term of years before they are granted their first leave, so that from the begining their roots strike deeper.

I am afraid that one could go on indefinitely with these rather invidious comparisons. They are certainly somewhat odious to our British *amour-propre*, but they are none the less intriguing. It would be so interesting to know, for instance, why the Dutchmen and Dutchwomen of Java use that most practical and economic steed, the 'push bike', on every possible occasion, while we English would rather die (or leave the country, 'broke') than be seen on one. It is painful to reflect that the only possible reason can be that the one community is sensible, and that the other is foolishly snobbish. . . .

The Dutch, those painstaking, orderly, methodical people—the race that a recent writer has described as sharing with the Scandinavian peoples the distinction of being the most civilized in present-day Europe—have made of Java, as nearly as may be, the perfect colony. It is a sort of tropical Holland, but a Holland modified and adapted to its new environment with all the skill that human brains could devise. . . .

Java is certainly a motorist's paradise. But there are angels at its well-barred gate, and each of them is thrice armed with a flaming sword.

The first is the Customs Officer, and his sword is a very flaming one indeed. The first time I took a car into Java, it was nearly seven hours, spent on the wharf in the Equatorial sunshine, before he lowered his weapon and let me pass, and even then I was scarcely over the threshold when even stronger forces were arrayed against me.

I arrived one exquisite morning by one of the 'KPM's' [Royal Dutch Shipping Line] admirable steamers, complete with car, and with a hundred dollars in Straits currency to cover the twelve per cent *ad valorem* duty required by the Customs,—a sum, by the way, which was a handsome compliment to my shabby, faithful old 'Chev', whose market value, I well knew, was far from being in proportion to my affection for it.

The ship tied up, the car was slung ashore with admirable ease and efficiency, my luggage was put in it, and off I drove

'Java is certainly a motorist's paradise.' (From J. Z. Van Dyck, *Garoet en Omstreken: Zwerftochten door der Preanger*, Batavia: G. Kolff and Co., 1922)

down the long line of busy wharves to the barrier, where I left the car, and went into the office to pay the duty and be 'cleared'.

I spoke no Dutch in those days, and having been warned that it is extremely tactless to address a Dutchman in Java in Malay (which he is apt to take as a suggestion that you suspect him of Eurasian antecedents), I started by asking the officer in charge if he spoke English. He assured me, not without a touch of asperity (for the Dutch are good linguists and proud of it), that he did. I thereupon explained my car was outside; would he be so kind as to assess it for duty?

He merely chalked my baggage without opening it, and then, after a pause, asked very courteously if there was anything further he could do. 'What ... Oh! ... I see ... a car' he said. 'You will find plenty outside'; and added, waving away the bundle of dollars I had proffered in token of my willingness to pay the duty, 'No ... no ... the fare is only four guilders.' Then, raising his voice, he called to the group of native taxi-drivers awaiting fares at the entrance, 'Heh ... the nyonya [lady] wants a car!'

But the nyonya didn't. This time I sat down firmly on the counter and, regardless of his feelings, explained in my very best Malay. Light dawned at last in the official eye. 'Ah ... Quite....' he said airily. 'Your own car. Why didn't you say so? Please give me the Import Pass.'

'Import Pass'? What on earth was that? I had never heard of it. But it seemed I ought to have got one from the Dutch Consul before sailing.

'No car can enter Java without an Import Pass,' said the official severely. 'You must hire a car, and leave yours here till I get the papers from Singapore.'

Now I had no mind whatever to depart in a hired car, leaving my own standing on the wharf for a week. It occurred to me that the right people to extricate me from the tangle were the firm who had shipped the 'Chev'. I learned that they had an office at the port, and set off (on foot in the blazing heat, for I was not allowed to take the car through the gate) to see them.

I found the office, and after long explanation the affair was duly located. Innumerable papers were filled in; every one was most kind, if a trifle slow. At last I was given an imposing document, all in Dutch, with which I walked gaily back to the Customs, confident that now I should soon be flying along the smooth wide road to Weltevreden [Batavia's European quarter], bound for a bath, lunch, and all the comforts of civilization.

I greeted mijnheer of the Customs like a brother, so pleased was I to think that soon I should have seen the last of him. But, to my horror, he only glowered at the paper I handed him. 'This is nothing,' he said coldly. 'I want the *Import Pass.* Go back and tell them.'

I was half-way down the steps before he finished, and I 'went back and told them' so eloquently that a smart young man was sent back to the Customs with me to fix the matter up. It would take only five minutes, they said.

What a flow of gutturals followed! What a turning over of folios and regulations! And then the Smart Young Man turned to me with the air of one in whom a brilliant and entirely novel idea has just been born. 'We have found it!' he announced triumphantly. 'You must pay twelve per cent of the car's value!'

What was the good of saying that I had tried to do that an hour or more ago? So I handed over my bundle of dollars in silence.

But it wasn't so simple as all that. Again they were waved away. 'The necessary papers must now be made out. It is a long affair. It may take all day.... You had better hire a car....'

I glared. Nothing would induce me to hire a car. I had one. My luggage was in it, and I intended to drive ashore in my car and no other.... They gave in, and the papers were promised in an hour.

I decided to go back to the ship, which promised more comfort than the car or the Customs shed at midday in these latitudes. So back I went, walking dispiritedly down the long, busy, scorching wharf, up which I had driven so gaily a couple of hours earlier. Back to the empty ship, where the Chinese 'boys', not unnaturally, regarded me with considerable suspicion. But the time passed and, punctual to the minute, came the Smart Young Man, waving three sheets of foolscap, covered with the finest script, which I signed where I was told, in blind, beautiful faith that all would now be well.

Back to the Customs we trudged, along those endless wharves, and once again I tendered my bundle of notes, looking now rather the worse for wear. 'Dollars? What was the good of dollars? You use guilders in Java. You must change them. Where? Why, at the Post Office, about half-way down the wharf!'

So to the Post Office I obediently repaired. But apparently that institution had never seen a dollar before. They looked at them with the utmost dislike. 'Possibly you could change them at Weltevreden.... You had better take a car....'

I fled. And then the Smart Young Man came to the rescue once more. He would change them in five minutes. 'Splendid!' I said. 'I'll sit in the car and wait for you, and then I shall get to town in time for lunch after all.' 'B-b-but,' he stammered, 'I have still to get the signature of the Head Controller. It is very far away. It may be many hours.... You had better take....'

It was the last straw. But nothing could move me now. 'If it takes many hours—or a week—or a month,' I said bitterly, 'I shall still be here when you come back.' I bought some bananas at a native stall, and climbed into the old 'Chev'. There was to be no bath, and no lunch; but this one refuge I could claim in an inhospitable country. I would stay in it till I died.

An hour passed. Two hours. I ate my bananas and longed for a drink. The sun blazed mercilessly down. I grew drowsily interested in the doings of the gatekeepers, in the jokes they cracked, the cigarettes they smoked, and the amazing expertness of their expectoration. I began to feel as though I had known them for years. I noted the instinctive politeness of the ever-passing natives, bound upon this, that, or the other of their lawful occasions, and rejoiced that I was not in the same predicament at Liverpool or Sydney. I was almost enjoying myself; at all events the thing had become a habit. I rather think I must have dozed ... for suddenly ... there was the Smart Young Man at my elbow.

'I tell you it's no good,' I said sleepily. 'I will *not* take a car....' He looked hurt, and handed me a slip of paper and the change—for all the world as though it had been only five minutes instead of half a day since the affair started.

I could scarcely believe it, but that really was the end—of that chapter. One more short interview with the Customs man (whom we found fast asleep) and at last I drove through the gate of paradise, thinking that all my troubles were over. But there was worse to come.

I had been warned that the hand of the Law was apt to be heavy upon the motorist caught driving without a 'rijbewijs' (driving licence). This would, however, be merely a formal matter in my case, I felt sure, for I had been driving cars for more years than I cared to remember, and had a whole assortment of licences of various countries to show to whom it might concern. So with a light heart I presented myself at the appointed place: the Parapatten Politie Kantoor.

The first snag was the discovery that Thursday (on which I happened to arrive) was an 'off' day; I must wait until the next morning. However, that was nothing. At the appointed time duly I returned, and it was not long before I realized that getting a driving licence in Java was no mere straightforward matter of demonstrating that you can drive a motor-car.

First, so the Dutch officer told me, I must go to the Post Office and get a form with a one-and-half guilder stamp, bring it back, and he would fill it in for me; then I must go and have two photographs taken and bring them back to him; then I must be 'examinated' (he spoke perfect English), and if I passed the test I must come back and apply for the rijbewijs in duplicate, and with that he went back to his desk.

Well, there it was. It sounded as though it would take a week, and I wanted to start up country the next day. But suddenly I had an inspiration: a dozen or more 'mata-mata' (native policemen) were sitting outside, apparently awaiting orders. There was no fear of offending them by speaking Malay, so, addressing one at random, I asked if he would come with me and show me where to get a photograph taken. He jumped into the car, obviously delighted, and off we set.

The Post Office was easy: the little mata-mata got the form in a mere twenty minutes or so. He knew a good place to get a 'portret', and directed me to a queer, low-browed little shop in a side-street, where I 'sat' to a small, kimonoed Japanese, while her two tiny, yellow scraps of children fixed me with an unblinking stare almost worse than that of the camera. That little lady could have given points to many London photographers. There was no fuss and no posing. She sat me down, smiled engagingly, focussed, trotted across and laid one soft little finger on my chin to press it gently aside, and trotted back to her camera. There followed the muffled click of the shutter, and she vanished into the dark room, whence she returned in due course with the negative in the hypo. dish for inspection.

It appeared excellent—distinctly flattering, I thought—and when it had been 'passed' by my friend of the police, who took entire charge of the affair, and impressed on her the urgent need for haste, she promised gravely that two prints should be ready in an hour: the price was three guilders.

At this the police escort broke in with indignation. 'Much too dear,' he told her. 'The nyonya could have a dozen copies for that! . . .' But she only looked at him very gravely, and said softly but firmly in her pretty Malay, 'You wish for them in one hour, yes? I will give them to you. The price is three guilders.' It looked as though an argument over a guilder might hold me up over the week-end, so I hastily laid down three shining coins, took the Law by the arm, and drove back to the Police Station.

Here there was a new development. The friendly Dutchman had vanished from his desk, and in his stead sat a tall, coal-black gentleman from the West Indies, in magnificent uniform and shining top boots, who told me, 'The other toean has taken your papers. He may not be back to-day.' It looked like a deadlock. But my little escort was determined to see me through, and poured out such a stream of eloquence concerning my many virtues and the extreme urgency of my affairs, that at last the dark gentleman, with a shrug, jerked his head in my direction, with, 'Get your car. I will come.'

I got it, and sat there, waiting. Presently out came His Dark Magnificence, chatting with a Dutch policeman and roaring with laughter over some joke which (perhaps happily) was lost on me. They entered the car, slammed the door, and continued their conversation without a word to me. 'Which way?' I asked; and thereafter, at every turn or crossing, I would inquire over my shoulder for directions, which they gave a trifle impatiently, as though it were rather rude of me to interrupt. We went round and round, through labyrinths of suburbs, through streets wide and narrow; it seemed as though we were going on all day, when suddenly there came a new order: '*Stop!*' We were in the very middle of a crowded street, so I disobeyed to the extent of manœuvring to the side, stopped, and waited.

Nothing happened. The conversation behind me flowed on. With deplorable rudeness I interrupted. 'What now?' I asked. A gust of laughter greeted me, followed by a long pause. Then: 'You drive much too fast!'

Now, in comparison with Batavia's 'flying taxis' (which fully justify their name), I had been crawling. But my heart sank like a lump of lead. Was I to be refused a licence after joy-riding these two loquacious policemen round the town for the best part of an hour? Obviously the soft answer was my only hope. I apologized abjectly. There was more laughter and then silence. 'Well?' I said nervously. 'Turn the car round,' came the order. I turned, and drove on—very slowly this time. '*Stop!*' I stopped. 'We will get out here.' 'And the licence?' I said, rather faintly. 'Oh yes . . . of course, the licence. Take this' (scribbling on and tearing out a leaf from his notebook) 'to the Kantoor. They will see to it.' And out they both got and disappeared through a gateway, without another word.

I was completely lost, but I found my way back at last to the Parapattan, and presented my talisman. Alas, no one was in the least interested. There was only one hope: it seemed the time had come for action. I abandoned all deference, and, raising my voice, spoke with chilly hauteur. 'Mijnheer—gave me this' (reading out the name of my scrap of paper) 'and said my licence was to be made out at once. Be so good as to see to it. I am in a hurry.'

It worked. Someone took the paper and the photographs (which I had called for on the way) into an inner office, whence there issued forthwith the quick rattle of a typewriter. Then I was led (again under police escort) to yet another room, where I was invited to press my thumb first on to an inky pad and then on to the space reserved for it on the licence (turning it from side to side so that none of my criminal tendencies should go unrecorded); then I was waved to a small tin wash-bowl hanging on the wall, where with a grimy bit of soap and still grimier towel I tried to remove all traces of the crime. Then came the final gesture: 'One guilder fifty, please'—and most thankfully I handed it over. Where else in all the world could you get so long and varied an entertainment for so low a price?

And so at last I had a rijbewijs, a green card stating that Mevrouw—, aged— but no, I shall not disclose that, and neither does the licence, for that matter!—was permitted to drive a motor-car in the Dutch East Indies from this time forward for the term of her natural life. Below was a portrait of feminine beauty whose anguished expression and staring eyes could scarcely fail to move the stoniest policeman's heart to compassion, however heinous a breach of the traffic regulations the original might have committed. And the final garniture was a blurred black smear, purporting to be the imprint of what I had hitherto always believed to be a rather shapely thumb.

However, it was just as well I had it. I had barely passed the outskirts of Batavia when a determined-looking Dutchman barred my passage. 'Your rijbewijs,' he said. I handed it over. He took it, looked first at the portrait and then at me, smiled (sadly, I thought), handed it back, and waved me on my way. . . .

The roads of Java must surely be some of the most delightful highways in the world; at all events, I know of none in any country to compare with them. There are (Heaven be praised!) none of those bare, characterless racing tracks with which we have defaced the fair countryside of England; but the surface of Java's main roads is as perfect as that of any 'arterial' horror, and even that of the by-roads is so excellent that if it could be reproduced on the main highways of Australia it would transcend the Commonwealth motorists' wildest dreams.

The main road from Batavia to Soerabaya, which was built in the ruthless days of Governor-General Daendels, at the cost of many thousands of native lives, is the backbone of a magnificent system that has been gradually developed, especially during the last twenty-five years, by the Dutch Works Department. There are more than two thousand miles of main motor roads, and considerably more than as much again of by-roads; the mileage is always increasing, and such gaps as there are in the system are being rapidly filled. The making of these roads in so mountainous a country has called for great engineering skill, and the grades are so easy that often it is only a pleasant freshness in the air which tells you that your car has climbed a couple of thousand feet.

From one end of the island to the other you pass through panorama after panorama of unsurpassable beauty and extraordinary interest, a living chart of varied production and industry.

Rice fields, thousands upon thousands of them, yet never two alike, spreading far across the flats, and climbing the hill-sides on watery stairways; shady plantations of rubber trees, their feet in the soft green cover crops; modestly small cassava (tapioca) fields; acres of ground-nuts; solemn, gloomy teak forests; miles of sugar-cane; kapok trees, with their queer, horizontal branches, wasting drifts of snow-white cotton on the ground; fields of sweet-scented citronella; tobacco; maize; and, high in the hills, tea, and the invaluable, insignificant-looking cinchona.

You can see all these things, and many more, from the Queen's highway, without ever leaving your car. You will pass an exquisite lake now and then, and cross bridges over rivers innumerable, hurrying down in their rocky beds, brown with the rich silt they are bringing to the rice fields. And ever in the background are the mountains, without which no Java landscape is complete.

Motorists owe much to the vigilance of the Java Motor Club, familiar to all who use the roads as the 'JMC'. This most efficient body not only sees that the roads are kept in perfect repair by the various local authorities, but surrounds the motorist with a care that is almost fatherly. There is not a sharp bend, nor a zigzag, nor a steep or dangerous hill in all those miles of mountainous country but is clearly indicated on a warning board at the roadside; there are so many of them that there is not a stretch of road of any length without one. And sometimes, when the occasion seems to call for it—when more than ordinary dangers lurk round the corner—these signs wax almost hysterical in their kindly anxiety for your safety, pouring out their whole vocabulary of hieroglyphics indicating hairpin bends and up and down hills, reminding one rather of an excited Frenchman. One of these signboards that I specially remember, on a particularly steep and tortuous road in West Java, seemed to have abandoned all attempt to specify the dangers ahead, and simply announced: 'To Tjawi, many—!' followed by a whole row of terrifying exclamatory warnings.

Another invaluable service rendered by the 'JMC' is the provision of a board bearing the name of each town or village, hung across the road from tree to tree, so that you always know, as in other countries so often you do not, exactly where you are.

On the country roads the traffic consists almost entirely of motors, and when you begin to meet pony sadoes and wagons it is a sure sign that you are approaching a town. There are many motor-bus services between all the towns, to which both the native and Chinese populations have taken very kindly. The buses are always crowded, and considering the great distances that can be travelled for a few cents, it is not surprising....

It is hard to describe the native traffic police in Java. They are simply too good to be true! They would look far more at home in the *Pirates of Penzance* than they do on point-duty in real life, even in Java. No one could possibly take them seriously, and apparently nobody tries; no European, that is. You will occasionally see one of them holding up a native pony driver or cyclist, but motorists seem to ignore them completely.

They are, however, very pleased with themselves, and their equipment is certainly impressive. They carry a large and intimidating truncheon, which they flourish in their more wakeful moments, and wear a short sword slung behind them. What does it matter that it is far too rusty and dirty ever to be drawn? Their uniform is of thick dark cloth, and they slump unceasingly from one foot to the other to ease the discomfort of heavy boots on feet accustomed to go bare. The picture is completed by a wide-brimmed, shady, brown straw hat, coquettishly turned up at one side, a product of the native industry at Tanggerang. Whether these brave fellows contribute at all to the keeping of the island's peace I cannot say. But they are unquestionably a most diverting addition to the variety show of the streets.

Harriet W. Ponder, *Java Pageant*, London: Seeley Service and Company, 1935, pp. 250–2, 255–8, 260–1, 263, 57–66, 285–7, 281.

In the late 1930s, Ponder discovered the trains.

Strangely enough, there were many modern residents of Java who knew nothing of the railways except from the distance, until the 'depression' transformed the commonplace convenience of yesterday into too costly luxuries and suggested, among other economies, the possibility of travelling by train instead of by car. In Java the price of petrol is more than doubled by the petrol tax; and when, in the case of British residents, this was aggravated by the fall of the pound sterling from its par value of twelve guilders or florins to less than eight, it was obvious that motoring could no longer be taken quite so lightheartedly for granted as of old.

But every cloud has its silver lining. And in this case it was the discovery that even after years of familiarity with Java its acquaintance yet remained to be made from an entirely novel point of view. And had it not been for the 'daling van de pond' (fall of the pound: a phrase that soon became hatefully familiar) it is probable that the unique panorama to be seen from the windows of the Java State Railways would have remained unknown to me for ever.

The names of the '*Eendaagsche*' (one day train) and the Java '*Nachtexpres*' (night express) had both become household words in Java since these much advertised trains *de luxe* had replaced the former slow, strictly daylight service. The latter restriction had been in deference to native engine drivers' reluctance to risk offending the spirits that walk in darkness; for a man may be a first class mechanic and driver and yet remain faithful to the beliefs of his forefathers. But education can achieve wonders; and now, so the posters of the *Staatsspoorwegen* proclaimed on all sides, travel on either the day or night express compared more than favourably with any in the world.

I was sceptical about the attractions of railway travel in Java. But needs must when the devil drives. I had survived long railway journeys in other hot climates, and reflected that at worst it could hardly be as bad as a summer trip across Queensland. And so one fresh delicious morning, just as the eastern sky was changing from silver to primrose, I took my place in a second class compartment of the famous '*Eendaagsche*', and saw the sun rise as we ran smoothly through the sweet scented gardens of the outer suburbs of Batavia.

I soon wasted no more regrets that for the first time I was not setting out across Java in the car. Far from it. The long airy compartment with the passageway down its centre was almost empty, and I had my end of it entirely to myself; so pleasant was it that a very few minutes served to convert me to this, to me, new mode of Java travel. I knew the Great Post Road from end to end, and hundreds of miles of it and other lovely Java highways and byways were far more familiar to me than any in England. But it became apparent at once that the panorama to be seen from this railway carriage was a revelation to one who had hitherto only travelled the country at the wheel of a motor car. No longer was I a part of the tangled pattern of life upon the roads, forced to thread my way among *sadoes* [carriages], *grobaks* [wagons], bicycles, cars, buses, handcarts, stray cattle, goats, dogs, fowls, ducks, buffaloes, and foot passengers, ever watchful to adjust my pace to their vagaries. All these terrestrial anxieties troubled me no more. I had removed to another plane, whence, godlike and unassailable, I could ride at my ease and survey the ever changing scene as we raced along, high above crowded roads and villages, houses and gay gardens, and the chequer board of ricefields, on embankments that seemed to have been designed for no other purpose than to provide the best of all possible observation points. . . .

The passing show to be seen from the carriage windows had been so enthralling that I had scarcely noticed the internal arrangements of this admirable train. I still had the long coach almost to myself, and so could stretch my legs or change my seat as often as I wished in search of shade when, in our shifting course among the mountains, the sun had made it too hot for comfort on one side or the other. There was no doubt that the train deserved all the praise that had been lavished upon it. Electric fans hummed steadily to keep the heat at bay. The split bamboo seats were as comfortable as they were pleasing to the eye, and the best for coolness and comfort that could possibly be designed for tropic use. Exploration further afield did certainly reveal one flaw, but not in the second class by which I was travelling; instead of the cool clean bamboo here, and also in the 'third', the seats in the first class had the very doubtful distinction of being upholstered in shiny black American cloth: with the lamentable result that many passengers who travelled 'first' paid the penalty of their exclusiveness by sticking to the seat, and emerging at their destination with indelible traces of select travel written on the nether portions of white drill trousers or silk summer frock!

The conductor of the train was a wizened, cheery little old native, a most companionable person, with none of the haughty airs of conductors on European expresses. He seemed to spend the entire journey wandering from one end of the long train to the other, wiping dust off seats, windows, and woodwork: an admirable idea, by the way, that might well be adopted on railways elsewhere. Of the meals served in the restaurant car I cannot speak at first hand; but the air of bland repletion on the faces of the passengers who passed through my compartment on the way back from lunch suggested that its standard had been well up to the general excellence of the train. At all events the iced coffee which, on the friendly old conductor's recommendation, I ordered many times throughout the day from the *djongos* [waiters] who made periodical pilgrimages through the train, was delicious. Served in a tall glass, in which tinkled a big silvery chunk of ice, they cost about twopence each.

My fellow passengers were few, but just enough to provide an occasional diversion. An elderly Javanese, obviously of the upper class, smartly dressed in European white suit and native *batik* headdress, sat opposite to me for a time, and conversed with animation on the affairs of Europe and China, regarding which he was at least as well informed as the average Englishman or Dutchman, if not better. On descending at his station he was met by two low bowing servants and greeted with the utmost deference by the native stationmaster, past whom he walked with a truly Oriental air of arrogance and climbed into a smart *sado* with a uniformed driver.

Two dignified grey haired Chinese ladies, immaculately clean and tidy in the becoming *sarong-kebaya* costume, were also my companions for a short time; and so, less pleasantly, was a fat Hollander, by all the tokens a 'drummer', smoking a most pungent cigar, who boarded the train at a Mid-Java station, followed by a skinny little porter loaded with four bulging and obviously very heavy portmanteaux. These safely stowed, the owner took out his purse and carefully counted out three copper cents (less than three farthings) into the man's hand, receiving in return a deep bow and an apparently grateful '*Terima kaseh banyak, Toean*' (Thank you very much, sir)....

The station and railway employees are all natives, but the differences in status among them are made obvious even to European eyes. Stationmasters appear at each stop, resplendent in freshly starched white uniform, red and gold peaked cap, and laced shoes: manifestly dressed up specially for the arrival of the '*Eendaagsche*', the great event of the day. This functionary's wand of office is a little bat made of split bamboo about the size of a ping pong racquet, painted white with a green centre, with which he signals to the driver of the train. Ticket collectors wear blue drill suits, peaked caps perched on the top of their *batik* headdress, which 'grows' on the male Javanese head almost as inseparably as his hair, and leather sandals: these last indicating a certain status upon which all those entitled to wear them are very insistent. Porters wear black jackets and shorts, and go barefoot; but they too wear a *batik* headdress.

Harriet W. Ponder, *Javanese Panorama*, London: Seeley Service and Company, n.d. [*c*.1943], pp. 222–3, 226–9.

23
Geoffrey Gorer Looks beyond the Colonial Gaze, 1935

'Now one of the chief disadvantages of living in hotels and cars', writes Geoffrey Gorer in the foreword to his Bali and Angkor, *'is that everything seen through their windows is likely to be* couleur de rose....'[1] *Gorer, a British-born esthete and scholar who specialized in national culture, found Java*

[1]Geoffrey Gorer, *Bali and Angkor, Or Looking at Life and Death*, Boston: Little, Brown, 1936, p. viii.

consistently less enthralling than earlier travellers, even the mythic Borobudur. The Dutch had been overpraised, he concluded, and imposed crushing taxes upon the people. Strikingly, where so many others had found Holland's indigenous subjects to be content and picturesque, Gorer encountered slums, squalor, poverty, and prostitution. The same Java?

FOR those, like myself, whose knowledge of scientific botany is negligible, Java offers few attractions other than volcanoes. For the Dutch tourist agencies the one thing foreigners really like is volcanoes; and in Java you can walk up to them, drive up to them, ride up to them, or fly over them. Since this inducement is so widely offered there must be some demand for it; but except for excessively Freudian reasons I cannot understand it; as far as I am concerned one volcano is as boring as the next one.

Round the volcanoes small beauty-spots have been preserved, one or two of them, such as Kawah Kamodjan [Kawah Kamojang] and the Dieng plateau, quite pleasant; otherwise as much as I saw of it Java's scenery is dull and monotonous. The country is so thickly populated that there are very few stretches without corrugated iron roofs, and the model cultivation is æsthetically uninteresting. Rice terraces give character to a landscape, but tea plantations look like a Victorian laurel shrubbery, extending over miles; the teak is an ugly tree with its big bilious leaves, and it is planted in geometrical forests; the young rubber trees have spindly trunks and dirty leaves. In Java as a whole there is no fresh colour.

There are four towns in Java which would be considered important in any European country; but it would be difficult to find in Europe four towns of as little interest. The old quarter of Batavia, with its canals and its red brick houses, has the picturesque qualities of the old and shabby; the ethnological museum there is not entirely uninteresting and there is a small aquarium in Batavia and a small zoo in Sourabaya which contain a few uncommon animals and fishes, including in Sourabaya adult gorillas. Otherwise these towns have as little interest, character, or beauty as Zenith, where Mynheer Babbitt sells real estate. There are, it is true, slums full of Javanese and Chinese as there is probably a negro quarter in Zenith; the climate and the bright clothes worn by the inhabitants prevent these parts being wholly depressing. But to get any pleasure out of such a spectacle you must consider people with a deeper pigmentation to be of another species than yourself, untouched by poverty and discomforts and squalor....

I did not personally find the Javanese very sympathetic; despite their fertility they give somehow the impression of being a race of old and exhausted people, only half alive. This impression may I think be due partly to their religion, and to the abysmal poverty of the greater number. Poverty, especially uncomplaining and involuntary poverty, is numbing and repulsive anywhere; and Mohammedanism is the most deadening of all creeds. A purely personal point which prevented me enjoying their company was the question of size; I do not like being among people who appear smaller and weaker than I am, unless they have corresponding superiority elsewhere; I dislike the company of those I feel to be my inferiors. In very early youth the Javanese boys are pretty and the girls sometimes very handsome and attractive; but such beauty is very transient. The poverty and unemployment has caused a great deal of prostitution on the part of both sexes; in the towns the more enterprising pursue you on bicycles.

Javanese art is a galvanised corpse. Such as it is, it is a relic of pre-Mohammedan Hinduism. It consists of performances of episodes from the Ramayana in a variety of media, through living actors, puppets, and shadows thrown on a screen. Except that in the last two cases the dialogue is recited by a single person there is little difference between the performances. They are all spoken in High Javanese, a language not one ordinary person in fifty properly understands; they are all stylised out of all relation with life; and the musical accompaniment seems to me to be without interest or invention.

The stylisation of movement for the actors requires very elaborate training, and has produced a few interesting poses and movements. The most novel of these is the stance for male dancers with the feet some way apart and pointing at right angles to the body; the knees are bent and the thighs are opened so that they lie in the same plane as the trunk; the arms are also held akimbo in the same plane, so that the man becomes an almost flat and excessively angular creature; this is as it were position one for male dancers, and most of the dancing is elaborations from this position. It is effective for some dramatic dancing, particularly fighting dances. Whereas the men exaggerate their bones and angles, women pretend they have none; women dancers approximate as nearly as they can to an upright snake or eel, or a boneless wonder. . . .

The puppet plays and shadow plays are even less interesting; the only pleasure to be derived from them being the ingenuity with which the dolls and silhouettes are manipulated. The dolls have, besides the usual hold, as in Punch and Judy shows, thin sticks attached to their wrists, so that the arms can be moved realistically; they have nearly as much life in their movements as the dancers. The silhouette figures, cut out of leather, have a certain weak rococo elegance. They are thrown on to a linen screen, lit from behind. There are some amusing stylisations of scenery in these performances, a pointed fan representing a mountain, and so on. . . .

At Djocjakarta [Yogyakarta] there was a 'modern' theatre, where under the old conventions modern plays, inspired by the more naïve films, were performed. I have never seen anywhere such dull and apathetic playing, not even in the West End of London. Boys took the women's roles.

Djocjakarta and its neighbourhood is the only portion of Java which to my mind repays the journey. Its chief interest, of course, lies in the ruins of the different Hindu and Buddhist temples; all the most important are within fifty miles of this town. Moreover, at Djocjakarta and Soerakarta [Surakarta] the work of colonisation has been carried out slightly less thoroughly than elsewhere; the old sultanates still remain with their full ceremonial and protocol, even though the powers of the sultans are extremely limited.

We received a printed permit to visit the kraton, or royal establishment, of Soerakarta. On this permit were printed a number of rules in three languages. It was forbidden to take photographs or notes in the kraton, to pass any remark about it or any feature of it in any language, to measure, or attempt to compute, or enquire as to its size, the number of its inhabitants, or any other detail, whether of the whole or a part. Visitors must dress as though they were visiting a European sovereign.

Wayang. (From J. Koning and D. Wouters, *Mooi en Nijver Insulinde*, Groningen: P. Noordhoff, *c.*1927)

We were very uncomfortably hot in our dark clothes when we arrived at the entrance at the indicated hour. Once there we had to get our permit signed and counter-signed; pass through a series of sentries, and then wait till an escort should arrive. We were shown over the buildings by one of the sultan's descendants, a youthful prince who was, without exception, the most beautiful person I have ever seen, languid and graceful as a Persian miniature. His complexion was the warm brown of some Spaniards, a trifle red on the cheeks, his slightly pouting lips almost crimson; his features were extremely regular, and his large dark eyes a perfect almond, fringed with incredibly long, curved lashes. His black hair grew low and irregular. He was very short and slight, but perfectly proportioned; like all the other male inhabitants of the kraton he had his breasts and arms bare; he was dressed in a sarong of brown batik, with a broad waistband of green velvet, which surrounded his body several times; in the band were fixed two krisses, one with a plain wooden handle, upright against his spine, the golden wood only a little lighter in texture than the skin it touched; the other, slung at his side, was handsome with silver and diamonds. He was very friendly, with an easy smile, and I should think as intelligent as a well-trained dog; what need has such a perfect animal of more?...

The object of the greatest single interest in Java is the ruined Buddhist stupa, or memorial shrine, the Borobodur. People even claim that it is the finest of all Indian buildings. Since I have not yet been to India I cannot tell whether this is true or not, but I sincerely hope not, for as a piece of architecture it is only equalled in insignificance and lack of interest by the Pyramids, and as a shape it is even uglier. It is a squat square pyramid, built round a hillock; each side has a double projection, that is to say that if each side were divided into six equal parts the second and fifth project over the first and sixth, and the third and fourth over the second and fifth. By this means the building appears from a little distance to be circular, and at the top the square is circled. The building consists of a series of galleries or terraces; above the base are four galleries of decreasing height sculptured on both sides, then three circular terraces dotted with small stupas, shaped like open-work bells with long handles, or cloches, a Buddha inside each lattice-work of stone; in the centre is the real shrine, a gigantic solid stupa or stone bell, surmounted by a huge octagonal column which was originally crowned with three stone umbrellas. The sides of the building also are covered with numberless Buddhas, each in a pinnacled niche; the whole effect is like a slightly squashed bowler-hat, bristling with spikes.

This symmetrical mess of stone is covered with the most lovely sculpture. Except for the numberless Buddhas, seated with folded legs and with the hands and arms in the (I think) ninety-two hieratic positions, and a few heraldic beasts by the doorways, it is all in low relief, mostly on sandstone, much worn by corrosion and exposure and difficult to see except in the right light. Round the four galleries, a long pilgrimage, the story of the Buddha in his different incarnations and that of the different Bodhisattvas are carved. The stories are often difficult to follow, but the grace, composition, and liveliness of the different panels—for it is all panelled—need no interpretation....

Hard by the Borobudur is a small rectangular shrine, the Mendoet [Mendut] temple. Its exterior, save for a handsome stairway, is unenterprising; the dark interior contains the most beautiful Buddha I know, one of the dozen or so masterpieces of sculpture in the world. The Buddha is seated on a throne, over which a drapery falls. He is naked except for a loin-cloth which falls between his legs. He is portrayed as a large and muscular, not particularly beautiful man. His feet are together, his knees separated; the elbows are a little above and farther out than the knees, the fingers joined in the attitude of teaching. The noble and contemplative head, with the eyes lowered, is covered with symmetrical and stylised curls, and the long ears are the only perpendicular lines in the statue. Behind the head a curved and pointed fan echoes and reverses the composition of the body, and the back of the throne, spreading like wings from the Buddha's shoulder, repeats on another plane the parallels of the base. I know of no other single piece of sculpture which so successfully combines every canon of composition into a whole of such complicated simplicity, so alive with dignity and holiness. Had I to choose an idol to pray to, this would be it.

Geoffrey Gorer, *Bali and Angkor, Or Looking at Life and Death*, Boston: Little, Brown, 1936, pp. 27–35, 37–40.

24
David Fairchild Views Java from the Air, 1940

Following his stay in Buitenzorg in 1896 (see Passage 15), David Fairchild embarked upon a career of plant exploration. Working mainly for the United States Department of Agriculture, he roamed the world in search of plants suitable for commercial cultivation in the United States (Egyptian cotton, for example, and the date palm). This work afforded him the opportunity to revisit Java, which he did at least twice, in 1926 and again in 1940. On the latter occasion, he sailed the seas of South-East Asia in a specially built junk, the Cheng Ho.

Garden Islands of the Great East *is his account of that memorable trip, during which he first assayed Java from the window of an airplane.*

AS the plane of the KLM cleared the suburbs of Soerabaya and the countryside was spread out below us, it was evident that Java was crowded with little homes. These were clustered together in kampongs, and clumps of tall feathery bamboos marked their whereabouts. I could see the garden patches, the bread-fruit and durian and mango trees near by, and the irrigating canals feeding the 'sawas', or rice fields with water from the mountainside. Where were the corrugated galvanized-iron roofs that stare up at one from other tropical landscapes? There were none; roofs of bamboo-shingles, thatch, or weathered tile were universal. Little children looked up at us as they played in the neatly swept dooryards. I could see the turtle-doves in their bamboo cages swung from bamboo poles stuck in the ground. I knew the walls of the houses were made by splitting bamboo stems into strips and weaving them together; that the floors were made of the same material; the drinking cups, the buckets, the brooms, the spoons, the forks, and every imaginable utensil in the house likewise, all from plants 'grown on the place', and that the boundary lines which divided the various households was some pattern of bamboo fencing.

As we climbed higher and the sawas reflected the light from the clouds, I could see how they descended, in an almost unbroken series of large flat terraces, from well up the hillsides to the seacoast. There, great ponds resembling the sawas took their places, ponds in which vast numbers of fish were raised for food—not so profitably, however, as I learned later. We were travelling over Middle Java, headed for our first stop, Semarang—a hundred miles away, over country so densely populated that from 250 to 500 people were living on each square mile. Yet in this area below us there is not shown on the map a town of more than a few thousand inhabitants. They do not live in towns, these people. The Javanese are skilled agriculturists who prefer the country. But their families come on so fast that the problem of food has become a serious one; one with which the Netherlands Government has long struggled.

I had hoped we would see at least one of the smoking volcanoes as we flew past, but although the Merapi was active, fleecy clouds shut it from view. Here and there were forest plantings of teak; tea and cinchona plantations; Hevea rubber patches belonging to the Javanese; and the larger, light green areas of sugar-cane, planted on lands rented by the big companies. If only the plane had been a blimp! I can imagine it would be fascinating to make a careful and leisurely study of Java from the air.

David Fairchild, *Garden Islands of the Great East*, New York: Charles Scribner's Sons, 1943, pp. 158–9.

25
Peter Kemp Arrives in Java at War's End, 1945

Following the Japanese surrender in August 1945, British and British Indian troops occupied Java in preparation for the return of the Dutch. To the same end, Peter Kemp, a British officer, was assigned to lead a reconnaissance party to nearby Bali and Lombok. Along the way, he passed through Java where Indonesian nationalists under Sukarno had already declared independence and where British forces were attempting to keep the revolution at bay. Java, as Kemp reveals in this passage, was no longer a safe place for colonizing Europeans.

BRIGHT sunlight, soft breezes and a blaze of tropical greenery was the picture I had formed of the islands of Indonesia, described by the Dutch writer Multatuli as 'a girdle of emerald around the Equator'. Reality, when I stood forlornly beside the sodden runway of Batavia airport at half past ten on the morning of 17th February, was grotesquely different. A thick, warm curtain of rain, falling from a blanket of puffy grey clouds and whipped across the airfield by the violence of the monsoon, blotted out the horizon and drained all colour from trees and grass and buildings. My fellow-passengers from the Dakota were driven away to their various units, leaving me standing in the open—bewildered, dripping and alone.

After a quarter of an hour an open jeep splashed up, driven by an apologetic young subaltern who took me to the Hotel des Indes, a large sprawling building once the queen of hotels in the Far East. War and the ensuing emergency had deprived it of its former extravagant splendour. Gone were the enormous meals of *rijsttafel*, each requiring a posse of fifteen waiters to carry the dishes, which had sent so many colonial administrators and business men to their early graves; and the high, airy suites with their wide verandas were stripped of most of their furniture and crowded with extra beds.

The situation in the capital was quiet but tense. Troops were required to carry arms in the streets, there was a strict curfew at midnight, and some quarters of the town and port were out of bounds; but the hours of daylight generally passed without incident. The nights were full of danger for the foolhardy; almost every morning patrols would retrieve from the drainage canals the dismembered bodies of British or Indian soldiers who had defied the curfew in search of liquor or women; girls indeed made a practice of enticing troops into their houses to be set upon by their menfolk.

The sudden end to the war brought about by the bombs on Hiroshima and Nagasaki found South-East Asia Command quite unprepared for the problems of policing and administration that immediately followed. For the Dutch East Indies, which had only been placed within the British sphere of operations by the Potsdam conference in July, there were scarcely any troops available. In Java the nationalists, encouraged by Japanese propaganda during the occupation and armed after the surrender by the Japanese Sixteenth Army on the orders of its commander, seized control of most of the country, proclaimed their independence and organized a provincial government under the leadership of Dr Sukarno. With their own demand for freedom they identified, illogically, their claim to sovereignity over all the peoples of Indonesia, many of whom differed widely in language, religion and culture from the Moslem Javanese, of whose influence they were bitterly resentful and afraid. The various nationalist groups, many of them little more than robber bands admitting no loyalty to the Sukarno government—they included the Communists under the Moscow-trained Tan Malaka—were united only in their hostility to the Dutch and to the British, who attempted to restore order in the Dutch name.

British and Indian reinforcements landed and, by the time I arrived, had secured perimeters defending the towns of Batavia, Buitenzorg, Bandung, Semarang and Surabaja. Only between Batavia and Buitenzorg was communication possible by road, and then only in convoys, which often had to fight their way through ambushes at a heavy cost in casualties. The nationalists did not spare their prisoners, whom they usually put to death by hewing in pieces to the greater glory of *Merdeka*, a risk which added to the hazards of a forced landing when flying over that jungle-covered, mountainous and enchantingly beautiful country. . . .

Formerly among the richest countries of Asia, with a happy and prosperous people, Java in 1946 [1945] presented a melancholy spectacle of neglected paddy-fields and derelict plantations abandoned by an impoverished and terrified peasantry.

Peter Kemp, *Alms for Oblivion*, London: Cassell, 1961, pp. 84–6.

26
S. J. Perelman Witnesses the End of an Era, 1949

While working on a series of travel articles for Holiday Magazine *in 1949, American humorist Sidney Joseph Perelman visited Java. As Perelman and his party arrived aboard the* Kochleffel, *the bitter struggle between Indonesian nationalists and the Dutch was drawing to a close and, as Perelman notes, 'the imperial goose was cooked forever'. In the passage below, Perelman captures the Dutch unflatteringly in the final moments of empire. Since the United States had refused to support the Dutch in their efforts to reclaim the Indies after the Second World War, Perelman, as an American, was subject to much anger and suspicion. (On an earlier trip to Asia, Perelman had been accompanied by the bearded illustrator Al Hirschfeld, who is mentioned in the passage.)*

ONCE, however, we managed to anesthetize ourselves to a few minor details like the food and the unrelenting xenophobia, shipboard life fell into a fairly pleasant pattern. There were always enough fresh excitements—flying fish, squalls, prahus, sea birds, and innumerable islands—to punctuate our indolent routine, and it was difficult to believe, the morning we anchored at Tandjong Priok, the harbor of Batavia, that the South China Sea lay behind us. Since the steamer was due to sail the next morning for Surabaya, our glimpse of Batavia was hardly more than fleeting. Through the good offices of a friend of Hirschfeld's—who, incidentally, had traversed this territory fifteen years before and left a trail of worthless chits—we were privileged to drive around the city, glut ourselves with dreamboat Chinese food, and visit the museum. Batavia, apart from the old fort and a sprinkling of seventeenth-century houses, offers approximately as much appeal to the senses as Poughkeepsie, except that it is hotter and devoid of Vassar girls; but the museum is a fit subject for dithyrambs, and were it not for the fact that a dithyrambectomy in childhood inhibits me from using them freely, I could unlimber some pretty lush superlatives.... There were in the capital, it should be noted, no visible signs of the blood-letting that was rife in the interior of Java, the daily ambush of Dutch convoys and the extinction of whole kampongs in reprisal. The Dutch, apparently impervious to world-wide censure of their invasion of the Republic of Indonesia, were currently pretending that their *coup de main* was successful and that everyone would be playing patty-cakes shortly. The truth was, nevertheless, that they controlled only a few isolated areas and those only by overwhelming weight of arms. You had merely to witness the sullen contempt with which the Javanese treated their white protectors to realize that the imperial goose was cooked forever—a dismaying fact from which Britain in Malaya and France in Indo-China were still girlishly hiding their eyes.

Soon after the *Kochleffel* cleared Tandjong Priok for Semarang, three new passengers made their appearance: Mr Chen, a young Chinese businessman en route, like ourselves, to the Moluccas, and a Mr and Mrs Hoogmeister. Mr Chen was one of those fantastic Celestials who, at twenty-four, have acquired the poise and commercial acumen of a merchant prince. He knew to a decimal what the zloty and the pengo were being quoted at in the free markets of Tangier and Bangkok, he had dealt in every outlandish commodity from gum copal to rhinoceros horn, and there was hardly a port from Tsingtao to Thursday Island he had not set foot in at one time or another. . . .

Mr and Mrs Hoogmeister, the other new arrivals, constituted no such social addition but were quite as remarkable. The former was a wizen-faced, Ichabod Cranelike exporter from Celebes, married a fortnight to a Frenchwoman thirty years his junior and uxorious to the point of folly. Though the lady was not downright misshapen, she was certainly nothing to heat the blood, and the surveillance he kept her under, his elaborate concern lest she stray out of his sight, brought to mind a Palais Royal farce. Madame Hoogmeister was seeing the Orient for the first time, her husband confided to me; he had fixed up what he described as a lavish home for her in Macassar, complete with a new Studebaker and a Frigidaire—had, in fact, gone to the devil's own expense to gratify her caprice, but it was pathetically plain that the bride was already fairly cheesed off on her surroundings. She kept complaining about the heat, the helicopter cockroaches which disrupted her sleep, the coarse and uninspired food. Hoogmeister was in a swivet; one's heart went out to him since he was, by his own admission, such a humane and lovable figure. When he talked about the Indonesians, whom he called 'my children', the hackles were assured plenty of exercise. He was fond of observing that in the good old times, the natives always crouched on the floor in his presence to show their subservience, but nowadays, emboldened by our American rubbish about democracy, they invariably stood up, as if they were his equals.

'Mind you, I'm a decent chap,' he protested. 'If a coolie asked me for a light for his cigarette, in a proper tone, I'd gladly give him one.' His wife cocked her head and regarded him affectionately.

'You really like these people, don't you, dear?' she inquired.

'Like them?' he repeated oracularly. 'I don't like them—I *love* them!'

Any expectation we entertained of visiting Semarang, we learned on reaching it, was fatuous; the Dutch military had proscribed the port as too dangerous for civilians to land. Consequently, the ship lay well out in the bay while its cargo was being lightered ashore through swarms of sharks and angelfish ceaselessly circling about for food....

It was somewhere between Semarang and Surabaya that I began to get an inkling of the friendly interest being taken in our party by the Dutch. Father de Groot had repeatedly tried to inveigle me into a political discussion, hoping I would reveal myself as a secret agent, but to no avail. One afternoon, the children bounced in with tidings that he had been grilling them about my opinions. Was it not true that their daddy held a Communist party card and deeply admired Russia?... Had they noticed any bearded individuals, redolent of vodka and caraway seeds, frequenting our home? The tots conducted themselves with textbook sang-froid. They replied that on the few occasions I came home, I was too befuddled with malt to talk intelligibly about anything, let alone politics. Having spent most of my life in jail, they continued, I was ineligible to vote. The only bearded visitor they recollected was a man named Hirschfeld, who drew amusing doodles on cardboard but otherwise appeared to have no grasp on reality. Father de Groot retired snarling to his breviary and thereafter made only one reference to any of us. He told Mr Chen that our children were much too precocious.

On Sunday noon, in a white-hot glare that made the pavements dance and shriveled the brain to a raisin, the four of us teetered down the main street of Surabaya toward its principal oasis, the Oranje Hotel. More leisurely and countrified than Batavia, and perceptibly cleaner, Java's second largest city also seemed far more colorful. The people strolling past represented every conceivable racial strain: pert Javanese youths in immaculate whites and Moslem caps, tall melancholy Hindus, Buginese boatmen, delectable Madurese ladies in bajus [shirts/blouses] and sarongs, and prosperous Chinese compradores, twirling cotton umbrellas, in striped pajamas and sola topi [pith helmet]. The particular sound that struck

Exotic Javanese women in a busy market place, by Carl Shreve. (From *Romance Calling*, KPM Line, *c.*1930)

the ear and set the rhythm of Surabaya was the measured, musical clack of teklets, the wooden clogs worn everywhere in the archipelago, but by nobody with as much élan as the Madurese charmers I refer to. Stooping to retrieve one that had fallen squarely in my path (a teklet, that is), I suddenly found myself swept full-tilt into the Oranje lobby by my wife.

'I know that gambit, Jack,' she said coldly. 'You're stalemated before you start. Now you freeze right there while I scare up some *rijsttafel*.'

S. J. Perelman, *Swiss Family Perelman*, New York: Simon and Schuster, 1950, pp. 78–80, 82–4, 86.

27
Harold Forster Finds His Place in the New Indonesia, 1952

When, in 1952, Harold Forster learned that he was being posted as a teacher to Java, he 'made a hasty round of the London bookshops to search for information about [his] future home—and came back almost empty-handed'.[1] *So, along with his wife Coral, he went to discover Java for himself. Gadjah Mada University, to which he was assigned, was barely up and running. Forster warmed to his new charges, but life in Yogya did have its limitations, as he tells us below. On holiday, the Forsters made a pilgrimage to Mount Bromo. Here they witnessed the feast of Kasada—held in memory of young Prince Kusuma who had offered himself to the volcano in repayment for his family's good fortune.*

AT last, through the heavy tropic darkness, our train came gasping into Jogja. That was how the name of our destination was spelt on the outside of the carriage, and

[1]Harold Forster, *Flowering Lotus*, New York: Longmans, Green and Co., 1958, p. vii.

that was what it was always called, though the correct form was Jogjakarta, or Djokjakarta, or even, in the full Javanese version, Ngajogjakarta, while Raffles in his *History of Java* (1817) spelt it Yug'ya Kerta. The popular abbreviation, informal and universal, was a sign of the place it held in the affections of the people—Jogja, the stubborn city of the Sultans, the sanctuary of Java's ancient traditions, the cradle of Indonesia's freedom.

At that moment, however, our main feeling was relief rather than romance, our interests practical rather than historical. All we wanted, as we climbed stiffly down to the station platform, was a good wash and a good meal. There were excuses for our somewhat mundane attitude. For one thing we were not tourists; we were to live and work in Jogja for at least three years. The amusing discomforts, the almost desirable difficulties that add a chiaroscuro to the tourist's memories, assume different proportions for the resident—there is no comfort in an infinite prospect of sweat and tepid rice. Then, again, we felt that we had been packed off with inconsiderate haste (through no fault of the Indonesians, it is true) on an unnecessarily risky journey. When we arrived at the port of Djakarta [Jakarta, formerly Batavia] the previous morning, I had expected a few days in the capital for consultation and acclimatisation, and then an easy two-hour flight to my post. The railways, our Dutch shipmates had assured us, were quite impossible, being regularly attacked by the bandits and not infrequently cut. It was a shock to be met at the dock with two tickets for the 6.40 train next morning.

In the circumstances we had not slept well. The heat of Djakarta is oppressive and our bungaloid hotel, built round a courtyard full of washing and only too near an odoriferous canal, was a sad contrast to the spotless spickness of our Dutch liner. As with most Indonesian hotel rooms there was a small open terrace attached, on which we sat as long as we could, watching the little house-lizards crawling up the walls and over the lamp-bowl, a situation which assured them of plenty of prey but also made their digestive processes transparent. At last we were driven inside by the mosquitoes and the droppings of the lizards. Five empty iron bedsteads, gloomily shrouded in dirty beige nets, crowded our bedroom but only emphasised our sense of abandonment. An inadequate electric fan buzzed and creaked on the washstand. It was almost a relief to be called before dawn, which took place about 6 a.m. with monotonous regularity all the year round....

Then came our train journey, nearly thirteen hours of it. The first-class carriages were spacious, comfortable, and at the start, clean. As soon as we began to move, however, the snag appeared. The engine ran on wood, its tender stacked high with logs like a timberyard; if we left the windows open, we were soon smutted with soft, unbrushable clots of wood-ash; if we shut the windows, we were boiled alive. This dilemma rather diminished our powers of appreciation as we rattled across the lush and steaming plains of the north coast to Tjirebon [Ceribon]. It was all new to us, the flooded ricefields glittering like broken mirrors, the stooping lines of coolie hats and the moon-shaped horns of the buffaloes, the coconut palms shooting skyward to burst into leaves like green rockets, the tattered banana groves and all the other trees whose names we did not know. But inside our moving oven thirst competed with sleep for our first interest. There was a restaurant service on the train, but we had been warned against this and supplied with a packet of inadequate sandwiches. Where the Dutch had stressed the danger of bandits, the Americans had impressed on us the perils of ice.

'You must never drink water in Indonesia unless it's been boiled', they warned us earnestly, 'and never, never touch their ice or you'll get dysentery for sure.'

At last our thirst became intolerable. We found we could order beer, bottled and so presumably safe. The bottle was warm when it came but in each glass stood a large lump of ice! It was too much—we decided to risk dysentery; and after that we drank water and ice freely all over Java without ill effects.

After the ice the bandits.... As we turned up into the hills after some six hours' journey in order to cross to the south coast, we were joined by an armed guard of soldiers. I am not particularly bandit-conscious; seven years in Greece just after the war taught me to regard them as a natural hazard of travel, a stroke of bad luck like a tyre-burst. But I always felt more anxious when provided with military protection. The Communists of the Greek hills were more likely to go for government troops and arms than to indulge in mere highway robbery: in Java the bandits were political too, though not of the same persuasion.... Our guard therefore increased my anxiety rather than allayed it, but on the whole the comparative cool of the hills more than compensated for such worries. As we climbed up over the backbone of Java, a long chain of volcanoes, the scenery became more varied and abrupt, yet essentially the same—everywhere a luxuriant green to the tops of the surrounding slopes. At Purwokerto we had reached the south side of the mountains and were safely through the worst bandit country. Everyone got out of the train to stretch and relax and smoke, and it was here that we first consciously noticed that clinging perfumed air that is so typical of places where a number of Javanese are gathered together. Soon we traced it to the fumes of the blunt cigarettes on every side; even little boys of six or seven were smoking happily and unrebuked as they squatted on the platform's edge. Surely it could not be opium? Fortunately we found an English-speaking fellow-passenger who informed us that these were the famous *kretek* cigarettes, in which the tobacco is mixed with cloves; their crackling gives the cigarettes their name.

The last stretch of two or three more hours through flat country seemed endless, leaving us as empty of romance as a squeezed lemon. It was well after dark again (dusk was as regular as dawn, between 6.30 and 7 p.m.) when we finally emerged from our carriage, stained and hungry, to be greeted by a smart young University official with a fine green Plymouth. Our troubles, we felt, were over as we swept into the wide front court of the Garuda Hotel with its sunken garden and its lily pond and the canopies over the stairways glowing with multi-coloured glass. A first-class room had long been booked for us, perhaps one of those upstairs with a balcony. The manager soon shattered our dreams. Unusually solid and podgy for an Indonesian, with a round cropped head, he complacently denied all knowledge of our booking. Documents, entreaties, threats left him unmoved—all the first-class rooms were occupied by important visitors for the celebrations and he could only offer us a second-class one at the back. At last we allowed ourselves to be led, still feebly protesting, to a kind of slum opposite the kitchen, an ill-lit cell with a vast plank bed and no bathing facilities. There was a bathroom nearby, reached by a corridor open to the public gaze; but it was a poor specimen of the 'Dutch bathroom', the only type to be found in Java and even at its best unsatisfactory to English prejudices. It consisted of a narrow tiled chamber with a stone tank of water in one corner, a small aluminium can with a handle across the top and sometimes a cold shower. This one had a shower, but it was rusty and cobwebbed and all the taps were waterless. I gazed with horror into the stagnant and slimy depths of the tank, which had obviously not been drained out for months, but at last I took the plunge—or rather (for we had been warned not to try to squeeze ourselves inside) I plunged the can into it and hurled the water desperately over myself. Feeling refreshed but not clean, we changed and went across to the dining-room.

We were to get to know this room only too well in the next ten months and slowly the service and even the quality of the food improved. But at the time of our first encounter it was at a very low ebb. A large gloomy hall with dark brown panelling, stained glass windows, and a peeling yellow ceiling chilled our anticipations. We wandered from table to table trying to find a cloth that was not stained and patched; a waiter, whose only uniform was a black sidecap of the sort that had become the symbol of nationalism, padded up in bare feet and dingy vest and planked down before us the whole meal together, complete from soup to fruit and all already cold—cold and watery consommé, cold and stodgy white rice, cold and damp brown vegetables, and the inevitable bananas.

Coral, who had faced so much that day, could not face this and burst into despairing tears.

Next day was our wedding anniversary and we rose as late and lazily as we could. Even so it was only 8.20 when we came out of breakfast, for a long lie-in was not really pleasant in such a climate. The heat roused you soon after dawn and though it lay heavily upon your eyelids and your will-power, the dank sweat of the pillow and the Nessus-touch of the mattress soon forced you to drag yourself to the shower. As we tackled the manager about a change of rooms, our University friend suddenly reappeared with his car.

'Are you ready?' he asked gaily.

'Ready for what?' I answered with some bewilderment.

'For the Commencement celebrations; they start at 8.30.'

In our flurry the previous night we had not thought to enquire about the celebrations mentioned by the manager nor imagined it possible they might concern us. But now it appeared that all those important visitors had come to attend a University function and I was expected to make my first appearance there. With blasphemous haste I changed from shorts into unseasonable *sub fusc* and the green car ploughed its way through seas of bicycles down a long straight street to the Sultan's palace. Through a white conventional gateway we came upon a wide and grassy square, across which we sped to the high-roofed forecourts of the *Kraton* with its slender pillars and elaborately decorated pediment of flowers and bees and snakes. Up a wide flight of stairs we hurried to a second pillared hall, crowded with dignitaries and students of both sexes. But I was diverted to a smaller building on the right, just in time to join a crocodile of learned persons, the leading figures swathed in black robes heavy with velvet. Their heads were crowned with tasselled academic caps in the shape of a lotus blossom, the emblem of the University but not, one must hope, symbolic of its spirit; thick silver chains made the Professors look even more like mayors in mourning. Ought I to have worn my own Cambridge robes? I had not thought to bring them to this tropic land where even the Jesuits wear white. Yet on second thoughts why should not the trappings of learning be just as important as the trappings of an army in creating the requisite spirit of pride and emulation? Gadjah Mada University might be only three years old, but it was determined to rival the solemn splendours of those Dutch foundations at which most of its Indonesian professors had graduated.

The crocodile wound through the packed hall and up onto a low marble dais, where a smaller roof below the main roof, both richly carved and gilded, formed a canopy over a polished grey plaque. Here the Sultan used to sit in state three times a year. Our procession divided respectfully on either side of it and settled in serried lecture-room chairs facing one another and the speaker's rostrum. The proceedings began.

For the next two hours I was able to sit back and take stock of the place and people among whom I was to work as a lecturer in English. . . .

My eyes wandered to the sea of eager brown faces on my

right, the students of the new Indonesia. They were dressed in Western fashion, the boys in open white shirts and trousers, the girls in simple cotton frocks—the price of modernism is the rejection of the picturesque. In the streets of Jogja we had seen plenty of both men and women clad in Javanese dress, but they looked to be mainly of the uneducated class; in this hall the only men to be seen in turbans and skirts were elderly officials of the Sultan's household. In the case of the married ladies, however, national costume was still in favour and the front rows, where the wives sat, were like a herbaceous border of sombre brown skirts and brightly flowered blouses. As I studied the students' features, in which the common denominators seemed to be straight jet-black hair, large and liquid black eyes, flat noses and wide, toothy and charming smiles, one of my cherished preconceptions about the Far East was exploded: here at least were no inscrutable and impassive Orientals but young people full of gaiety and restlessness, whose free and open expressions presented no mask to the world—a world where a mask was no longer needed. Their complexions varied from a rich brown, sometimes with a delicate bluish bloom like a plum, to a lovely golden tan, such as our pallid socialites vainly pursue in long and expensive sojourns on the Riviera; but with normal human cussedness they in their turn admired and strove after our enviable Northern pallor....

Whether from reaction or the cumulative oppression of the heat, lassitude set in quickly and I found myself, after only a few days in the glamorous East, deep in ennui. This sudden spiritual and physical inertia was a recurrent feature of the next ten months, for it was not till we escaped from the hotel that I began to feel that I was getting nearer to the real life of Java. It was a malaise compounded of exhaustion, frustration, loneliness, discomfort, lack of vitamins and a kind of pervasive hysteria that magnified each petty irritation into a grievance and made life mountainous with molehills. The Hotel Garuda—formerly Grand Hotel de Djokja—had already got me down.

It was not entirely the fault, one must admit, of Bagong. His European guests were a difficult lot, riddled with complexes that might have puzzled an expert psychiatrist. But even elementary Western psychology seemed to be beyond him. His manner, sometimes offhand and indifferent, sometimes jocular and familiar, was calculated to redouble the annoyance of those who approached him with some well-founded complaint about the food or the laundry, the service or the water-supply. He was one of the few Indonesians we met who did not display a natural and delicate courtesy—and the state of the hotel at that time demanded the highest tact on his part. As it was we suspected the worst: we saw deliberate malice in his failure to improve the conditions; we denounced him as a secret anti-white; we were goaded to fury by his ultimate, desperate evasions.

'I am sorry,' he would smile when we came with some cast-iron complaint, 'but I am only the under-manager. You must ask the manager about that.'

'Then please send for the manager.'

'He is not here; he lives in the town.'

'Well, get him on the telephone, then.'

'I am afraid he is away in Djakarta just now.'

'When will he be back?'

'I do not know.'

He had all the technique of the Orient in dealing with Western importunity, except its polish.

The hotel was the Quartier Latin of Jogja—an involuntary and disgruntled Quartier Latin. Owing to the housing shortage most of the rooms were occupied by foreign experts, who formed a hard core of unwillingly permanent residents. Dutch, German, French, Italians, Austrians, Hungarians, Yugoslavs, Russians, Indians, South Africans, English—one or two of each, plus three or four Americans; doctors, technicians, teachers, business men, musicians, anthropologists, chemists: these formed a group incoherent both in interests and in means of expression. Too various in origin to form a genuine community, too different from our surroundings to escape into local society, we found common ground only in our grievances and frustrations. At our daily forays in the dining-room or seeking a breeze in the garden court at dusk we swopped atrocities, adding fuel to each other's fires and denouncing Indonesia with all the vehemence of thwarted love.

We were impatient, and the East demands patience. It was a period of painful readjustment, and unfortunately some of our Western friends did not stay long enough to readjust. At best, hotel life is unnatural, and we were living a life that belonged neither to East nor to West. Though the stream of Javanese life roared past our windows day and night, we were somehow isolated and apart on an artificial island, in a far-from-ivory tower. Cut off from the restraints of normal society, eccentricities flourished and passions flared till we became as odd as the inmates of any Grand Hotel in fiction. It only needed a shaping imagination and a dramatic climax to become a ready-made novel; but the climax always failed, the expected bangs subsided to whimpers, and our stories ended as a collection of disjointed anecdotes.

The Americans, of course, were the most impatient and it was they who consequently lost their balance most quickly in the slow-motion world of Indonesian bureaucracy. The turnover in United States personnel during our stay was tremendous in proportion to their numbers. They came eagerly, overflowing with goodwill, were shocked to be met with reserve and distrust, grew disillusioned and begged for instant withdrawal. Their desire for escape took different forms—Homer took to misanthropy, Hank to indiscriminate love; Al took to the bottle and bombarded the fountain with the empties, Wilbur to pills and potions for his thirty-seven allergies. The Europeans on the whole were less extreme but more pathetic. Theirs was a long-term despair, the gradual disintegration of the exile, for many had no home to return to, no country to own them. They were the gloomy drinkers, the vacant starers, the whisperers, the suspicious and touchy cadgers. The Dutch covered up their bitterness with a dour stolidity or a loud and beery heartiness.

Only the old Austrian, still as ramrod-straight as in his days in the Imperial Hussars, seemed to have found the answer in Yoga and the films of Fred Astaire. Every dawn we could see his legs waving above his balcony wall as he stood religiously on his head; and most evenings he would urge upon us the wonders of the latest American musical—ach! the precision of the dance-movements, the beauty of the breath-control! We found his detachment enviable but not imitable. The next slug in the salad or lost letter or scorched shirt would set our nerves jangling again like the bells in the street outside....

A walk down Malioboro [Yogya's main street] could be full of excitement. Alternatively you might try to cross Malioboro. At any hour between 6 a.m. and 9 p.m. this was a hazardous attempt, even at the official crossings. Cycles are less lethal than cars but more unpredictable, and the overwhelming majority of Malioboro's traffic consisted of bicycles—or tricycles. This was one legacy of Dutch culture at least that was unlikely to be cast out in a hurry. Nearly everyone seemed to boast a bicycle, down to the raggedest worker, and nearly every bicycle seemed to boast a passenger. But, though popular, cycling had not become vulgar in Indonesia. It was still an affair of high handlebars and upright figure, a dignified gliding, and there was no sign yet of the strenuous crouch and bulging buttocks that disfigure the English weekend countryside. Even passenger-riding was a refined art. The carrier-borne female sat side-saddle, balancing herself at an elegant incline, her hands modestly restraining her billowing skirt and never, under any circumstances, clinging to her escort. To touch a member of the opposite sex, even a husband, in public was considered shocking, or as they termed it themselves, ridiculous. Many a raised eyebrow, many an embarrassed giggle, was caused by our walking down the street arm in arm.

The tricycle is the taxi of the Far East. The Indonesian type, known as *betja*, differs from its cousins in Singapore and Bangkok, for the passengers ride in front and the driver behind. Some said that the Japanese insisted on this arrangement during their occupation, as their fastidious noses were offended by the wafts from the sweating coolie in front. Whatever the reason, there were other advantages of this system that more than outweighed the perils of head-on collision. The passengers enjoyed more privacy than in Singapore, where the driver sits alongside, and a wider view than in Siam, where the foreground is filled with the coolie's background. Though in bad weather the cramped seat, the low folding hood and the dank and dingy rain-curtain flapping against your knees were apt to induce claustrophobia, on a warm evening and a smooth road the *betja* provided a rare combination of pleasures, coolness, speed, colour, and even a kind of music.

The padded seat, slung between the front wheels, was

gaily upholstered in red or green or blue leather and the mudguards decorated with flowery patterns and still more flowery names. Sometimes the wooden back-panel displayed a primitive willow-pattern landscape or a volcano à la Hokusai. But in Jogja we never saw the wild efflorescence of uninhibited folk-art that was common in Surabaya, where the rash of decoration spread even behind the wheels and the seat itself was a riot of popular mythology: dragons and cowboys, elephants and aeroplanes, tigers and lovebirds, Mickey Mice and, not least, girls—girls at the mirror, girls on motorbikes, girls in swimsuits, girls retiring behind the bed-curtains with a coy 'Goodnight!' *Betja* music, however, was as delicate as *betja* art was crude. A shrill and ghostly humming, like the song of the wind in the pines, rose and fell with the speed of the machine. Only after many months did I trace the source of this unearthly accompaniment, a rubber band stretched tightly between two small bars under the seat and working on the principle of the Aeolian harp. . . .

It was hardly to be expected in the circumstances that everything [at the university] would be running smoothly. The lack of trained and experienced administrative staff in particular was glaring, as it was in every branch of Indonesian government. The serried desks of the offices on the verandahs of Pagelaran were a wilderness of bureaucracy; urgent papers would take weeks to pass down a single line of In and Out trays. It was ten months before my appointment was officially confirmed by the Ministry, and until then they could only offer me an 'advance of pay'! In the strained atmosphere of mutual commiseration at the hotel, without pay, without housing, without books or equipment, without vitamins, and without apparent prospects of ever getting any of these, the little band of Westerners swore again and again that they would quit. Yet somehow most of us carried on, for there was one important item of which we all had a plentiful supply—students. They had conquered so many difficulties to get where they were; it was up to us to face a few ourselves for their sake.

The mere physical problem of finding decent quarters for such a rapidly growing flood (463 in November 1949; 3,439 at the time of my coming in 1952; and 8,570 when I left in 1956) was formidable enough. Accommodation was the biggest headache and first priority. Thanks to the Sultan the various Faculties at least had roofs (often very ornate ones) over their heads, and classes could be carried on, albeit in a somewhat amateurish manner. But with the target figure of 10,000 undergraduates being rapidly approached, Jogja was bursting at the seams. Somehow, somewhere they were all fitted in; packed into bungalows, servants' compounds, garages, cubicled *pendopos* [verandahs, or open-air halls], or when approved, *kampongs*. But it could not go on like that. . . .

I found the students much more easy to get on with than their elders, my colleagues and contemporaries. Kindly and courteous as the older generation was, I never felt that I had really got close to them. It was not merely their natural reserve—and Jogja was noted even among the Javanese for the aloofness of its people—but the sense of a lingering, perhaps unconscious, but irradicable mistrust of the European that stood between us. This made me hesitant to assert my ideas and cautious over such mild reforms as introducing some written examinations; for they were almost morbidly sensitive to the least suggestion of a 'colonial' attitude. With the younger generation, however, I felt no such inhibitions. Independence had come to them young enough to be natural, before they could be conditioned to an inferiority complex. They had sufficient self-confidence to accept guidance. As one of them wrote in the University magazine, complaining of the lack of foreign professors:

'At present a common disease seems to prevail in Indonesia, the fear of foreign influence. . . . This attitude gives rise to an exaggerated prejudice against Westerners in general and the Dutch in particular. In our opinion such a fear is unfounded. It will only make us feel inferior to them. . . . Our students are old enough not to adopt the bad influences. We always like to say "We are not a weak nation." A strong nation is not so easily persuaded to give up its principles.'

With my students I felt free to express my views and give advice frankly, dispensing with the extra wrappings of tact and patience that made dealing with my colleagues so laborious and frustrating. I like to believe they knew I was on their side. . . .

Gadjah Mada University was only one aspect of the immense effort of the Indonesian government to raise the people from 90 per cent illiteracy to modern educational standards, with a ten-year plan for universal elementary education by 1960. Ambitious and probably impracticable as this plan was, it nevertheless made for one of the most impressive and successful undertakings of the young state. For the whole people was behind it. The thirst for learning, so long denied them, was intense, as notable in the adults as in the streams of tiny brown dolls to be seen in every village, padding happily along the muddy pathways between the bamboo groves and the canals, proudly clutching their exercise books or slates.

But the first essential for schooling is a sufficiency of teachers, and so every possible candidate was quickly given employment in the schools. As English was now the second language of Indonesia, taught for a total of six years in the Junior and Senior secondary schools, there was an urgent call for anyone with knowledge of the language and many of my students, as soon as they passed their first-year examination, were engaged to pass on their inadequate knowledge in the local schools. It was a desperate position, for owing to their school duties they could not give full time to their university studies; engaged in examining others, they had no chance to prepare for their own examinations. Before long they and their pupils were fellow-undergraduates, and it was amusing to observe how quick was the return on my work, as my teaching was reproduced to me at second hand. Yet gradually the standard was rising and gradually the senior students managed to pass out, the first B.A. in English being granted in October 1954.

The difficult years were far from being ended, the standard was far from satisfactory, the syllabus far from perfect, and the staff still utterly inadequate—consisting as it did of myself alone. But something was being achieved, and it was a privilege to have a share in it....

Kasada night sounded so romantic that we determined to try to see it one year despite the uncertainties of transport, lodging and dates in Java. It meant an expedition of three or four days and we planned to make the most of it by travelling via the mountain-lake of Sarangan and the historic sites along the Brantas valley. The journey started disastrously. At Madiun the car that was to take us up to Sarangan was requisitioned by an unexpected Minister and the cost of a taxi was so fabulous that we preferred the discomfort of a bus. The bus, cramped and rickety, stopped short of the lake and storm clouds forced us to spend the night as the sole occupants of a resthouse in the unpronounceable village of Ngerong. At dawn next day we scrambled up the hill on ponies (which looked a little less dangerous than the rough sedan chairs swaying on four coolies' shoulders), for a quick look at the misty lake cupped on the mountainside, before hastening back to Ngerong to catch the morning bus to town. Alas, no bus appeared for hours and by the time we were down again in Madiun we had missed our train to Blitar. . . .

The train that eventually took us to Blitar had nothing but second-class carriages of such a vintage that most of the window shutters were immovable and we had to sit in grimy semi-darkness. When darkness fell outside as well, the whole coach was served by only one small oil lamp. Here and there the evening was lit by the dance of myriads of fireflies over the glimmering ricefields; and here and there the hard flare of a pressure lamp showed up one corner of some wayside platform. . . .

It was late evening when we at last reached Blitar, but from that moment our troubles ceased. This was entirely due to the kindness of the Regent, whose daughter was one of my students. A large car swept us to his residence, built in traditional style with a wide *pendopo*, marble-paved halls and high bare rooms. After supper he showed us round and pointed out a huge heavily carved table.

'They call that the Holy Table,' he said, 'because it has saved men's lives. In 1919 our volcano, Kelud, suddenly erupted and killed fifty thousand people. The lava rushed right in at the back door of this house—you can see the mark of its height on the doorpost there—and the then Regent and his family only saved themselves by climbing on top of this table. It is very solid, you see.'

'But did they have no warning?'

'Kelud never gives warning, that is why it is so dangerous. There is a lake in the crater and nobody can tell when the weight of the water will make it collapse. It has broken out three times already in this century.'

'So it might erupt at any time—even tonight?'

'Perhaps; the last time was in 1951, but it only killed some cows. And after 1919 they built a special refuge out there in the garden.'

The refuge was comfortingly close, built on a high stone platform; and we slept soundly after our long day.

Next day the Regent's daughter took us to the ruins of Tjandi Penataran, the biggest Hindu temple in East Java....

Penataran seemed doubly serene in the clear cool sunlight of late afternoon. Nor was the atmosphere of peace too deceptive, for the temple dated from one of Java's most peaceful periods, the height of Modjopahit [Majapahit]. Smooth lawns and carved grey stones against a backcloth of coconut palms made a pleasing general picture and closer inspection added artistic interest. In the middle stood a small restored temple with a steeply stacked roof; behind it another, notable for the great snakes undulating round the eaves; and behind that again the main temple rose in three terraces with wide platforms and relief-covered walls. The top was flat with a pit in the centre; but whether this was for royal ashes or a holy tree, or indeed whether Penataran at all resembled the earlier mausolea like Prambanan, was not clear. Its ground plan reminded us more of a modern Balinese temple than of the highly symmetrical pattern of Prambanan or Sewu, and there was also something Balinese in the flat *wayang*-like figures and vegetable whorls that filled the panels of reliefs. Penataran was typical of the later style of Hindu-Javanese architecture, when the rococo Javanese element prevailed, and it formed a clear link between the older monuments and modern Bali.

Not far away we came upon the so-called Pool of the Nymphs, which was really a pair of ancient stone bath-chambers in a deep hollow. Our approach disturbed some naked soapy youths at their evening ablutions under the flow of the monster-headed spouts. Gazing at these elegant showers with their permanent running water, where the Javanese had bathed perhaps every evening for the last six centuries, we could not help comparing the plumbing of Modjopahit with the rusty waterless apparatus at our hotel....

The Regent's kindness did not stop at hospitality. He confirmed the date of Kasada night, booked us a room at the nearest resthouse to the volcano and lent us the indispensable jeep to reach it. So we drove off up the circling course of the Brantas and, perhaps because it was lovely weather and a holiday, the East Java countryside looked gayer and more colourful to me than the flat green sameness around Jogja. Deep ravines and steep hillsides gave variety to the scenery; the roads streamed with shining jingling dogcarts, gallant with banners and feathers; and the files of women in bright plain *kebayas* and scarlet shawls, proudly upright under loads that they carried Bali-wise on their heads, made a pleasing contrast to the dingy browns and bent backs of Mid-Java.

A few miles past the pleasant hill-town of Malang we turned down a side road in search of Tjandi Singosari, founded as a mausoleum for the proud king Kertonegoro. On we bumped till we came upon two guardian giants, sunk deep in the road and so colossal that two boys were playing on one gigantic hand. But they must have guarded some other shrine, for we found we had missed Singosari temple behind a thin screen of trees some way back. An old crone conducted us around and regaled us with fluent but dubious archaeology. Comparatively modest in size, the building looked slightly unfinished owing to the unwonted plainness of the monster faces over the four doorways. The cells were empty except for one, where a typical wise man with flowing beard and protruding belly was identified by our guide as the fabulous smith, Mpu Gandring, maker of the magic *kris* that played such a part in Singosari's brief and bloody history. Some other images were arrayed in the grounds, but I could hardly believe that a stout negroid figure squatting on his haunches portrayed King Kertonegoro himself, while Raden Widjoyo, the founder of Modjopahit, was disappointingly represented by his lower half alone. . . .

At midnight the barrier at Ngadisari was lifted and we plunged into a maelstrom of bucking jeeps, skidding motorcycles, shying horses, goggled Chinese and villagers loaded with live chickens, palm wreaths and even a *gamelan*, all pressing eargerly along the narrow muddy track. At last we emerged on a cliff-top where the brilliant full moon showed us a vast basin surrounded by sheer cliffs and in the middle two great grooved cones that were hills in themselves. This was the famous Sand Sea, so called from the mist that gathers in the basin, making it look, in moonlight or at dawn, like a wide silvery lake. Down a precipitous track our jeep slithered into the mists. In the blind and featureles sands of the flat floor we quickly got lost and it took some searching to find the foot of the lower, wider cone, Bromo itself.

Up on the volcano's summit torches and pan-pipes and weird shrieks like Redskin war-whoops made the moonlit scene still eerier. We joined a stream of people toiling up the damp sandy slopes, somewhat hampered by the need to be wrapped in a blanket. From inside we could already hear a roaring like the sea. At length Coral stopped, defeated by the slippery steep, though our guide offered to drag her up with a rope. But I persisted and five minutes' stiff clambering brought me to the narrow rim, three feet wide and uncomfortably crowded. Now the roar was thunderous like the rumble of a train in a long tunnel, and I could gaze down to the cavern of red-hot lava far below. But despite the menacing growl and the boiling vapour clouds the inner slopes were alive with small boys shouting and rushing about to catch the offerings of coins, cigarettes and sometimes live chickens that the pilgrims threw from the rim; while down near the lava a wandering torch indicated that a *dukun* was conducting the real sacrifice there.

Even on the volcano's edge it was cold. But we learnt how bitter the tropics can be when we returned to our jeep to wait till we should be allowed to go back along the one-way track—at 8 a.m. The chill mist curled round us and all our thick clothes, blankets, a flask of whisky and even a Highland Fling hardly kept us from freezing. At last, soon after 5 a.m., the light crept over the cliffs and the cones went through a dramatic succession of hues from purple to green. Bromo's ridged slopes grew quiet and deserted again, and by six o'clock we could wait no longer. In defiance of the regulations we fled from the frosty sands and dissolving vapours, luckily meeting no incoming traffic.

As we drove down the deep wooded valley towards the golden sheet of the sea, we began to thaw out and discuss our experiences. Coral told me how, as she sat waiting for me, she noticed a young guide moving from group to group of tourists with a pair of shoes in his hand. She pointed him out to a shoeless Chinese beside her and he gratefully reclaimed his footwear, which he had taken off to make a safer descent. In the crowds and darkness he had lost the guide carrying his shoes and never expected to see either again. It was good to find that even today the pristine virtues persisted in the unworldly realm of ... Bromo.

Harold Forster, *Flowering Lotus*, New York: Longmans, Green and Co., 1958, pp. 3–10, 20–2, 28–9, 45, 50–1, 53–4, 239–47.

28
Arthur Goodfriend Scouts Java for the Free World, 1958

Arthur Goodfriend, something of a China-hand and a professional writer who worked off and on for the US State Department, was a passionate booster of the Free World during the heyday of the Cold War. In books like If You Were Born in Russia *and* What is America? *he extolled the advantages of the American way of life. The fate of many Asian countries, he believed, lay ultimately in the hands of their vast peasant populations. How could Asian villagers, he wondered, be persuaded to cast their lot with the West and to reject the seductive alternative of Communism? To find out, he arranged to spend a year in Java, close to the rice roots. He, his wife Eadie, and their children, Arthur and Jill, settled in Solo. The passages below recount the beginnings of their stay and introduce us to Goodfriend's indispensable go-between and interpreter, Samiek.*

WE left the *President Cleveland* in Hong Kong and flew to Djakarta on a Dutch KLM plane. The craft was immaculate and expertly handled. The stewardess was blond, pretty and attentive. She showered the children with *hopjes* [coffee-flavoured toffee] and treated the rest of us to a splendid Dutch lunch. I relaxed in the air-conditioned comfort and recalled my first visit to Djakarta, long before the war. It was Batavia then, the proud capital of the Netherlands East Indies. I remembered the strictly policed traffic, the carefully manicured gardens, the deferential servants; above all, the magnificent hospitality of the Hotel des Indes, justly famed as the finest hostel in the East.

Fifteen years later, on my second visit, the city had changed far more than in name. The decrepit buildings, the cracked pavements, the jerry-built slums that had once been lovely parks bore no resemblance to the fastidious colonial show place. The once-orderly boulevards were a tangled snarl of battered cars and bicycles. Homeless hordes of beggars and peddlers used the sidewalks as spittoons. From the street, the Hotel des Indes looked as gracious as ever, its dance floor and lounges gay with pink-faced men and women quaffing Heineken's beer. Inside, like all Djakarta, it was a shambles. Its disrepair, according to rumor, was aided and abetted by its Dutch proprietors to shame the Republic, and to make its most ardent well-wishers yearn for the dear, dead Dutch days they, the proprietors, hoped were not beyond recall. For the children's sake, I prayed Djakarta had improved.

The courtesy with which we were ushered through customs deceived me into thinking that it had. Once outside the airport, my hopes sank. A Communist Party sign bade us welcome. Beside it stretched a huge banner emblazoned with one word, 'Irian!'[1] A ragged child tugged at my sleeve for alms. Crimson betel-nut juice gushed from a passerby's mouth and spattered Arthur's shoes. He insisted his feet were bleeding and hid his face in Eadie's skirt. The light of anticipation in Jill's eyes was washed away by tears.

Our travel agent finally appeared. A genial Dutchman, he explained that some stupid native clerk had mislaid my wire from Singapore advising his office of our arrival. He warned of trouble at the des Indes. It was packed to the rafters, he said, with Russians recently arrived to set up their new embassy and, he added, 'sooner or later to take over the country'. Our baggage aboard, he guided his car into the turbulent traffic streaming toward the city. Pedestrians and cyclists responded slowly to his horn. 'Imbeciles,' he swore. 'The whole country is going to hell!'

[1]Referring to Irian Jaya, or West New Guinea, which the Dutch government had not yet relinquished to Indonesia. It would do so in 1962.

Little outside the car window disputed him. We drove through a complex of shanty towns, gashed by canals used openly and indiscriminately for defecation and drinking. Buildings and walls wilted in the heat and bore, either stenciled or scrawled, the one word, 'Irian!' The old Harmonie Club, once the temple of Dutch planters and traders, slid by, a sad, soiled shell. Beyond lay the des Indes's pretentious driveway.

A surly clerk registered us. A porter led us to two cells in a distant tier. A toilet and bath, shared by thirty guests, were a flight below. Their condition discouraged use except under duress. Eadie bathed the children in a sink and put them to bed. There were no mosquito bars. Bloody smears on the walls told of nocturnal battles waged by previous tenants.

The heat discouraged sleep. Eadie and I went down to the lounge for a drink. A trio of musicians pumped out fox trots and waltzes. White and Chinese couples thronged the dance floor. A familiar face rose from a distant table and beckoned. It was an old Embassy friend, who invited us to join his party. Consisting of the manager of an American oil field in Sumatra, an English sugar refiner from Surabaja, a Dutch banker, and the correspondent for a newspaper in New York, it was a gay group until someone asked our mission in Indonesia. My answer induced a heavy silence, broken at last by the American oil man. Sarcasm tinged his tone.

'My wife and I have been here nineteen years. All that time we've had one house boy. We raised him like one of the family. But when he goes back to his kampong, he might just as well be off to the moon. As to what goes on inside his head, we haven't a clue. If you ever discover what makes an Asian tick, do let me know.'

A sigh escaped the sugar man. 'I felt sorry for the blokes when I came out here. Raised the cane pickers' wages the first chance I could. Next day no one showed for work. We lost a lot of sugar. Company asks us to treat them like human beings—better conditions, inducements, that sort of thing. Nothing works the way it does with normal people. I've given up trying to figure the devils out.'

My Embassy friend shook his head sadly. 'Frankly, we've never tried to understand the rank and file of Indonesians.

We deal exclusively with the big wheels and count on them to reach the masses. Once, when the politicians and the people were united by the urge for freedom, the method worked. Today, there's no longer one ideal to bind the high and low together. The leaders and intellectuals have lost contact with the masses. And so have we.'...

Next morning the official in the Ministry of Information mulled over my request for advice on where to work.

'Indonesia,' he said at last, in excellent English, 'is big. It has two hundred ethnic groups and forty languages. You couldn't possibly know them all. But more than half of our eighty million people live here in Java. In central Java our culture is purest, yet it is adapting daily to changing times. My suggestion is, settle somewhere around here.' His finger circled the map around Surakarta. 'Work out into the surrounding villages. Then follow some of the transmigrants from the area to Sumatra, where they are carving new lives in the jungle. That way you can encompass the past, present and future of our country.

'Surakarta,' he continued, 'is the center of the universe. A tower has stood there for centuries to mark the spot on which the world revolves. Recently it burned down. Next day the monsoons broke early. Spirits, weeping at the tower's loss, caused the rain. The city has some of our finest theaters and music, and just the right combination of farming and light industry to give you insight into our economic problems. Another thing, Surakarta's women are famed for their beauty and passion.'

I explained that my wife and children would be with me. 'In that case,' said the official, 'housing will be a consideration. Every inch is occupied, but it happens that the palace of one of the local princes has been converted into a small hostel. It is our way of forcing our aristocrats to invest their gold in useful projects. It stands close to the center of town, right across from the Sriwedari, a pleasure park where wajang, or shadow plays, are performed. To know our people, you must know our wajang. In Surakarta you will eat wajang, drink wajang, sleep wajang. It goes on night and day.'...

Eadie and I talked it over. Despite the beauty and passion of its women, the Center of the Universe, she agreed, sounded perfect for our purpose....

The plane circled the flats around Djakarta and pointed inland toward the mountains, some of them rice-terraced almost to their summits, verdant, the water in the paddies shimmering in the sunshine. The central range mastered after an hour's flight, the southern coast came into view. We descended to Maguwo airport, a brown speck on the green carpet of central Java. Switching to a car, we sped along a good highway, past fields of rice, sugar, cassava and corn. In the distance smoked Merapi volcano, fertilizing the land with its ash. Here and there rose the ruins of lacy Hindu temples, the soil about them cultivated to their crumbling foundations by farmers pushing plows almost as old as the mossy stones. Joy returned to the children's faces and to Eadie's heart and mine. This was Java, here was the land, these were the people we had come so far to know.

Beside the Solo River, Java's largest and loveliest, sprawled the city. Our driver said people called the city Solo too. We entered between two monumental pillars and drove down the main street, named the Slamet Rijadi after a gallant revolutionary fighter, and bordered by Dutch bungalows converted into government offices, barracks and schools. A railroad shared the street with horse-drawn surreys, tricycle taxis called *betchas*, barefoot pedestrians and herds of goats and geese. Venders of food, toys, eyeglasses, magic potions, haircuts and horoscopes squatted along the footpath. Two movie theaters dominated the main intersection, their flamboyant billboards advertising an American gangland thriller and a propaganda film from Peiping. Run-down and architecturally undistinguished, Solo nevertheless looked alive and gay; the only note of sorrow was cast by the rubble of buildings wrecked during the revolution.

The Prince's palace, where we were to live, faced the Slamet Rijadi. It consisted of a rather elegant central building bordered on three sides by connecting cottages, each with its own garden and porch. Outside it resembled a Miami motel, but the reception hall and dining room, with eight carved pillars supporting an intricately sculptured ceiling, and our bathroom, with its hip-high tub of cool water, were Javanese. We introduced Jill and Arthur to their first Indonesian bath, filling a gourd with water and splashing them thoroughly. They loved not having to worry about wet floors and walls, and we all emerged clean, happy and hungry.

In the dining room an old man, barefoot and garbed in a bright sarong, set before us bowls of rice, curried vegetables, and bananas fried in coconut oil. The children sampled the fare and almost retched. From an adjoining table a tow-headed family wafted us sympathetic smiles. The man came over and introduced himself as Dr Heintz. He was a boyish, crew-cropped Wehrmacht veteran from Leipzig who, to escape the Communists, had wangled a three-year contract with a local hospital. His wife, Helena, plumply pretty, had eyebrows which conveyed her emotions better than did her halting English. At the moment their sharp angle expressed hunger. With their two boys, about Jill's and Arthur's ages, the Heintzes comprised Solo's entire foreign population. Our arrival swelled it to eight.

Dr Heintz said the absentee Prince's hospitality made few concessions to Western palates and local shops stocked little with which to augment the royal table. Until recently there had been bread of sorts, but government restrictions on flour imports had cut off this staple from their diet. Helena said the only escape from starvation was the Malang, a Chinese restaurant where, several evenings a week, they gorged on prawns, pigeon and other delicatessen. The Malang was to see us often. That evening, however, Jill and Arthur met hunger for the first time. It was to become a familiar companion in the months ahead.

I lost no time, next morning, getting down to work. My greatest need was an interpreter. From Heintz and others I learned of a man who spoke low, middle and high Javanese and English. Son of a hadji—one who had made the pilgrimage to Mecca—and himself a good Moslem, he had studied during Dutch times in a Catholic school, had edited the local newspaper, and now served as agent for the Garuda airways, an undemanding job. His name was Samiek, meaning breast-sucker.

Samiek proved to be a tawny Falstaff. From his round waist up, he wore Western clothes except for the black velvet cap that is Indonesia's national headgear. His lower extremities were swathed in a flowered sarong. His dark eyes were brightly shrewd, his voice loud, his laugh hearty. I liked him on sight and came abruptly to the point of my business. Starting with the birth of a baby, I told him, I wanted to witness everything that happened to an infant, a child, an adult throughout an entire life cycle. I wanted to visit schools, mosques, clubs, labor unions, theaters and family celebrations. I wanted to talk with farmers, workers, students, teachers, merchants and civil servants. Everything that happened to people, I said, was important, from the lullaby that mothers sang to their babies, to the services beside the grave. I asked Samiek to tell me frankly if he thought this could be done.

'Ten Indian boats come into the harbor', said Samiek cryptically, 'but the dog's tail stays between his legs. Some people shy away from what is strange. If you are one who will condemn or mock our ways, you will learn nothing.'

I assured him I was without prejudice, fully aware that though his people's customs might differ from mine they deserved as much respect.

'My people,' said Samiek, 'are like frogs under a coconut shell. Since the shell is all they have known, they think it is the universe. And so, since all the white men they have known have been Dutchmen, *belandas*, you will be a *belanda* too. They will fear you like a tiger that hides its claws. It will take time before they will speak frankly in your presence.'

'Where I live,' I said, 'people also believe in knowing a man's character before they trust him. Anyway, let us try. I will need all of your time for many months. We must arrange a schedule. We must settle the matter of your pay.'

Samiek rose. 'A sarong,' he said, 'is used as a protection against nettles. Money is used as protection against shame. I have another engagement. Please excuse me.'

Samiek's addiction to proverbs added spice to his speech, but left me uncertain of his meaning. His parting words made me suspect that my brusque reference to time schedules and payment had offended him. When days passed and he did not return, I was sure. Idleness was irksome and, on my own, I started to explore the city. Following the *rijadi*, I strolled past a stone prison and the post office to the Susuhunan's palace, fronting on a broad, dusty square shaded by holy trees and dominated by two medieval cannon converted into shrines. The labyrinth of narrow streets surrounding the square hid many mosques, a Chinese temple, the vegetable, meat and bird markets, the pawnshops and innumerable stores. From the poorly paved streets grassy lanes wandered off into the kampongs where dwelt the majority of Solo's three hundred thousand citizens. Each a little ward of about a hundred families, they seemed much like the villages on the road from Maguwo, pressed tightly together. Huts of split-bamboo matting stood in the shade of banana and palm trees, each on its own brown patch of beaten earth fringed with fruit-bearing bushes, and each filled with children and chickens. The air was pungent with the sweet smell of clove tobacco and the heavier smoke of charcoal fires oozing from the windowless walls. Narrow paths, bordered by open sewers, led to other kampongs, spaced farther and farther apart as they ranged outward from the heart of the city until, imperceptibly, town and country merged.

In one of these kampongs I passed a meat vender's stall and watched while he spitted tidbits of goatflesh on bamboo sticks, dipped them in hot sauces, and roasted a half-dozen sticks at a time over hot coals. Famished, I squatted between a *betcha* driver and a farmer laden with chicken-filled baskets for the market. The vender gave the sticks a final twirl and offered them to me. I pointed to my neighbors. The farmer accepted them and smiled. The *betcha* boy said, '*Belanda.*' I shook my head and said 'American.' He shrugged. In the dirt I sketched the stars and stripes. Both men looked and repeated, '*Belanda.*'

My meat was ready and I dug my teeth into the hot morsels. The others watched and laughed. The *betcha* boy said, '*Tidah belanda.*' Another man stooped beside us. It was Samiek. 'They say you are not a Dutchman,' he said. 'You eat *sate* just like they do. They ask which you like best, heart, liver or flesh?' I said I liked them all. '*Bagus,*' they said, 'good!' We all shared another round of *sate* with Samiek. Then, rubbing the grease off our fingers in the grass, we rose and shook hands. Samiek walked home with me.

'A torch,' he said, 'is useless to a blind man. I shall hold up the light. Whether or not you see is up to you. Soon a neighbor's wife in my kampong will give birth. You cannot use a rice thresher to pull a thorn. You cannot buy your way to the house of my neighbor with money. But we Javanese like to have pictures of ourselves, our wives, our children. Perhaps, if you will present him with pictures of the birthday celebration, you will be welcome.'

Never, thereafter, did I try to pin down Samiek to work schedules and rates of pay. Nor, to any Indonesians, did I ever again mention money. Not that such practical matters were overlooked. It was just that there was a right way and there was a wrong way to go about them.

The dirt floor of Samiek's neighbor's house was covered with a reed mat, set with banana leaves piled with rice and the meat, offal and eye of a buffalo. Male relatives and friends sat cross-legged along the walls. A *modin*—a layman learned in the Koran—mumbled a prayer. Behind a bamboo partition, women clustered about the mother-to-be, a child in her early teens named Rahaju. She lay upon newspapers spread over the earth. A *dukun*, or witch doctor, knelt beside her. As the rhythm of labor began, the *dukun* poured water into a bowl and added turmeric. The water turned yellow and she dropped in a brightly polished coin. Samiek explained that the 'gold water' was a disinfectant and astringent for mother and child. The coin betokened wealth.

Djogo, the young husband, a sandal cobbler by trade, was enthralled by my camera. When the natal moment approached, he insisted that I photograph the entrance of his first-born into the world. He sat on the floor and cradled Rahaju's head in his arms. The *dukun* gently pressed the girl's abdomen and arranged a sarong around her outspread limbs. She prayed to the good and evil spirits which, Samiek said, invade houses at this critical time. Djogo began to blow on Rahaju's head, gently at first, harder as the spasms quickened. The *dukun* crooned, 'Give it to me, mother. Come out, baby. Come out quickly, little one. Bring your brother, the afterbirth, with you. The gold water is ready to wash you.'

The girl responded silently to the *dukun*'s coaching. 'Rest now. Breathe easily. Now push hard and bite your hair. Allah be praised! It is coming! Here it is! Give thanks to Allah! It is a boy!'

The *dukun* rubbed the umbilical cord, then cut it with the sharp edge of a piece of bamboo. Blood was rubbed on the infant's lips 'to keep them always red'. The *dukun* dipped the baby, a lusty six-pounder, in the gold water, wrapped it in a clean cloth, and placed it by the mother. Fully conscious, Rahaju smiled down upon it through happy tears.

Djogo held a clay pot to catch the afterbirth. Rice, salt, flowers and perfume were added, followed by a coin, a pencil, a sheet of paper with Arabic characters. Samiek whispered that the placenta was the baby's younger brother who would watch after his well-being from the spirit world. The coin invited wealth, the pencil knowledge; the Arabic characters signified ability to read the Koran. We followed Djogo, who carried the pot in one hand, an umbrella in the other, to a hole dug beside the door. There he buried it and reverently covered the grave with flowers.

Within the house, the sarong was stripped off Rahaju, never to be used again except to cover the child in times of illness. The *modin* struck three sharp blows near the infant's head, to make sure it would not grow up to be nervous or lacking in poise. Then he whispered into his left ear: 'God the Almighty, I swear there is no other God than Allah. I testify that Mohammed is His messenger. Come, we shall pray. Praying is more important than sleeping. Allah the Almighty, we beg good fortune. There is no other God than Allah.'

The guests divided the food, wrapped their portions into banana-leaf parcels and, wishing Djogo well, left for their homes. Each reminded Samiek not to forget their pictures and invited me and my camera to be present at events impending in their families.

I was elated at the success of Samiek's stratagem, but also disturbed. I felt that my presence during the more intimate moments of Rahaju's childbirth had been an intrusion. Moreover, the camera had taken so much of my time that my notes had suffered. It seemed to me we needed a photographer on our team. Samiek produced a gaunt, raffish character named Bambang. From a Dutch news photographer to whom he had apprenticed himself, Bambang had acquired not only mastery of the lens, but other tricks as well. He handled his Speed Graphic like a pistol, bullied his subjects unmercifully, and behaved generally in the exaggerated manner Hollywood deems appropriate to his trade. Samiek, of course, had a proverb to fit him: 'If the teacher urinates while walking, the pupil urinates while he runs.'

Between them, Samiek and Bambang could transform my most improbable requests into action. They knew everything happening within a ten-mile radius of the city. Here a girl was being 'circumcised'. There a thief was being tried. In a distant kampong, a man was divorcing his wife. In another, a farmers' co-operative was meeting, or the Communists were staging a rally. None of these events was announced in my daily newspaper. No telephones connected Solo with these obscure villages. The pair clearly had sources of information and means of communication of their own. I counted on time to solve these and other mysteries that baffled Eadie and me.

One of the oddest of these was Samiek's habit of never being content with a gift. When Eadie learned that he was married, she gave him a small brooch for his wife. He astonished her by asking whether she had any others. She hadn't, and Samiek looked pained. On another occasion, when I bought flowers for Eadie, I presented Samiek with an identical bouquet to take home. He stared at it almost in dismay, as though it were in some respect deficient. Whatever we gave him, we invariably were left with the feeling that he was reproachful rather than grateful. In a more mercenary man, such behavior would have been less strange. It didn't fit Samiek.

Neither did his moodiness on Fridays. A garrulous, laughter-loving companion throughout the week, on this one day he was always glum and fidgety. On Fridays, too, Bambang's hand shook and his pictures came out blurred. Neither would volunteer an explanation.

Of all Samiek's idiosyncrasies, the one that puzzled us most was his behavior at mealtimes. Never a punctual man—'In Indonesia,' he'd say, 'even time is made of rubber'—he always displayed deep anxiety about the hour at noon and sundown. Murmuring that his wife was waiting, he would hastily leave. Eadie and I could understand his hunger, which we shared, but never the fact that he always took off in a different direction. We became more and more curious about Mrs Samiek.

Bambang was also married but he flirted atrociously with every pretty girl he met, apparently with great success. He enjoyed being teased about his conquests, laughing, looking knowingly at Samiek and saying nothing. We added this to our growing list of enigmas.

Only once did we quarrel. I wanted complete photographic coverage of a wajang performance at the Sriwedari. The mayor of Solo, in charge of such matters, granted permission with reluctance. We laid careful plans to exploit our opportunity. Samiek would translate the dialogue word by word. Bambang would move about below the stage, keeping one eye on the action, the other on me. Whenever I wanted a picture, I'd raise my hand. It was a foolproof arrangement except that, in the middle of the last act, Bambang ran out of flashbulbs.

Samiek quipped that Bambang's arms were too short to embrace the mountain. Bambang shrugged the matter off. I scolded them both for being so irresponsible and wasting the long evening's work. When my tirade ended, Samiek stood up. 'When we do well, we smell like flowers. When we fail, we smell like excrement. When your nose decides how we really smell, let us know.'

Shame overcame me when the mayor sent a note saying the play would be rescheduled so that his friend Bambang could photograph the final act. I searched all over town for the pair.

'Forgive me,' I said. 'You are flowers, now and forever.'

Samiek said, 'Americans can solve all life's secrets, but they do not know that in a thousand years nothing matters.' Then they laughed. Thereafter, I kept a tight reign on my temper.

As I had hoped, time and propinquity cleared up some of the mystery cloaking Samiek and Bambang. Blue Fridays, for instance. It turned out that Solo folk believe that evil spirits called *demits* attack people in bed on Thursday nights. Everyone visited or walked the streets until dawn, making up the lost sleep on Friday, the Moslem Sabbath. Only Samiek and Bambang, trying to appease both the *demits* and me, suffered. As much for my sake as theirs, I changed our day of rest to Friday.

Bambang's sexual ardor turned out to be a matter of politics rather than glands. 'For as long as anyone can remember,' he finally admitted, 'every Indonesian learned that the Dutchman was a better man than he—in the Army, the schools, the civil service. There was only one way and place that we could prove we were better than the *belandas*. That was in bed. When you tease me about my passion, I laugh. I am not laughing at your joke, but at the Dutchmen.'

None of these oddities struck me as anything more than amusing until, one day, an Englishman arrived in Solo. Smythe was with a large European tobacco concern which manufactured cigarettes for the Indonesian market. When imports were cut off, the local tobacco had become important. Smythe, convinced that Java's soil would produce fine Kentucky tobacco, had given seedlings free to any farmers who would take them. He had also taught them the best cultivation and fire-curing techniques. A fine crop had resulted. Now he was going out to pay the growers a surprise visit and look into the marketing. I accepted his invitation to go along. It seemed a rare chance to see foreign enterprise at its most enlightened.

It turned out at the first village that our arrival was expected. About twenty farmers were waiting, the local Communist leader at their head. Their manner was sullen, and I suddenly remembered it was Friday. The Communist spokesman, a lean, intense man, spurned Smythe's hand and announced that the company was cheating the farmers. Smythe calmly asked how.

'The company's prices are too low,' said the other, and the farmers growled agreement. Smythe drew a copy of a contract from his brief case and began to read its terms relative to prices.

A farmer interrupted. 'We cannot read your contract. It is men we trust, not written words. You promised us a higher price for Kentucky than for our *krossok* tobacco. Instead, we are getting less.'

Smythe's face reddened. 'Naturally,' he said. 'We paid high prices early in the harvest. Now with so much Kentucky on the market, prices are bound to be lower. That's the law of supply and demand.'

The Communist said, 'We understand but one law, Allah's.' Smythe's temper frayed. 'How dare you, a Bolshevik, talk about Allah!' He waved the contract in front of the farmers' faces. 'This is the tobacco code! This is what we promised! This is what we pay!'

A farmer answered. 'This is our country. We have our own code. You gave us the seedlings to place us in your debt. You promised many things but hid your true meaning in printed words we cannot read. Now, instead of payment, we get excuses. This is what comes of trusting a *belanda*.'

Back in the car, Smythe mopped his forehead. 'Bloody silly of me to lose my temper. How can you do business with people like that?'

I admitted I didn't know and told him of my own errors in dealing with Samiek and Bambang. 'Maybe,' I said, 'we ought to know more about their outlook on business, their ethics, their terms.'

Smythe seemed dubious. 'But I say,' he exclaimed, 'I'd like to meet your Samiek. Maybe he can tell me how the beggars knew I was coming!'

Samiek was nowhere about. Eadie said that though it was Friday he had worked with her on newspaper translations all afternoon. Then, suddenly uneasy, he had left just before supper.

I slept fitfully that night, disturbed by the day's misadventure. I had seen the heavy Kentucky bursting from the soil, and the scrawny patches of native *krossok*. Smythe deserved credit for every leaf. What a paradox that the Communist should be the farmers' champion, and Smythe be cast as their foe. I measured the gulf of ignorance that divided me from these people—even from Samiek, whom I should certainly know by this time. Why, I wondered, was Samiek so disturbed whenever he left us? Where did he go? Who was the mysterious Mrs Samiek? Why were our gifts never enough?

The answers came months later.

'As a Moslem,' Samiek said, 'I am allowed four wives. I have taken full advantage of my privilege. But each is jealous of the others, and I keep them far apart, in different kampongs. Allah help me if I forget with which woman I am to share a meal or bed. Their tongues have no bones. Should one receive a gift, the others know of it before it is out of my hand. Women!' He spat out the word as though it were bitter. 'They are blind to an elephant on their eyelids. But a louse they can observe on the other side of the sea!'

Arthur Goodfriend, *Rice Roots*, New York: Simon and Schuster, 1958, pp. 23–41. Reprinted with the permission of Arthur Goodfriend.

29
Frank and Helen Schreider Search for the Indies of Old, 1963

Frank and Helen Schreider were fans of arduous travel and, in Indonesia as in Latin America on an earlier trip, they moved about in an amphibious jeep dubbed Tortuga. This permitted them to get off the beaten path and served as a magnet for local people, who swarmed to see the miraculous jeep that was also a boat. On assignment for National Geographic *magazine (and accompanied by a large German shepherd dog), the Schreiders set out to find in Java and the other Indonesian islands something of the mystery and magic of the past—as it had been described to them by a nostalgic old Dutchman in Peru. The Java they encountered was considerably more complex, and dangerous. In the following passages, the Schreiders learn the ways of the Indonesian bureaucracy, and take us out in the Tortuga—to Bogor (formerly Buitenzorg), Bandung, and Ceribon. In the final passage, Frank Schreider recalls a violent encounter with one of Java's nascent strongmen, a member of the ubiquitous, powerful army.*

IF only we could have found from the start the Indonesia of the old Dutchman we had met in Callao.

How long ago that seemed, five years it was. At the time we were nearing the end of a journey by amphibious Jeep over the Pan American Highway. It was such a casual meeting; I can't even remember the old Dutchman's name, and intent on reaching Tierra del Fuego, the things he told us we soon forgot. But later we remembered our meeting with the old Dutchman.

It was at a little outdoor café near the water in Callao. It was August, winter in Peru; the sky was a dull slate and I remember that the sea gulls were like snowflakes in the air. All the tables were full and the old Dutchman asked us to join him. He asked about Tortuga, our amphibious Jeep, and we told him how she had carried us through storms and reefs in the Caribbean, the jungles of Mexico and the surf off Costa Rica to bridge the gaps in the Pan American Highway. A distant look came to his eyes.

'I would have given anything to have had a vehicle like yours when I was in the Indies,' he said.

That started it and for the rest of the afternoon we sipped Pisco Sours and listened while he talked of his life in the Indies. I remember that it was always the Indies he spoke of, never Indonesia.

Before the war he had been a planter on Java. 'Not a large planter, mind you, just a small one. A few hectares of tea, that's all.' During off seasons he used to travel, and once he went to the Spice Islands. 'You know,' he said, 'it was a shorter route to the Spice Islands that Columbus was looking for when he discovered your part of the world.' He told us of the Dragon Lizards of Komodo and the colored lakes of Flores. On Sumatra he saw elephant and tiger and rhinoceros tramping unhindered through rich groves of pepper, and Lubu tribesmen who lived in trees and hunted with blowpipes within miles of modern oil installations. He knew the head-hunters of Borneo where longhouses were decorated with polished and yellowed skulls. He wandered through the terraces of the great Borobudur and watched the life of Buddha unfold before him in stone relief. On the green cool nights, with the rice fields silver in the moonlight, he used to sit under a banyan tree on Java and listen to the music of the gamelan or watch a shadow play. 'And Bali,' he sighed, 'Bali is another world.' He had gotten out of Java when the Japanese occupied the Indies in 1942 and he knew he would never return. 'Just a few hectares of tea,' he said again, 'that's all. But it was all I ever needed or wanted. That was paradise on earth.' And I remember that he looked down at his drink to hide the tears in his eyes.

At the time we had no idea that that meeting would launch us on a journey along a three-thousand-mile chain of islands, through seas that Magellan, Drake and Bligh had known. It was only when our Pan-American venture was over and we were back home in California and felt the emptiness that follows the realization of any long-anticipated journey that we remembered the old Dutchman. Full of confidence in Tortuga that had carried us from island to island in the Caribbean and across South America's Strait of Magellan, we saw in Indonesia with her thousands of islands and hundreds of straits a challenge at least equal to the Pan American

Highway. We converted another World War II surplus amphibious Jeep into Tortuga II, a rolling, floating home, a redesigned and improved model of the one that had carried us on our Pan-American adventure. With the backing of *National Geographic* we gathered our equipment and booked passage to Asia.

The Indonesia we found when we arrived was not the Indonesia of the old Dutchman. The country was in a state of war and emergency. The Indonesians were blaming the Dutch; the Dutch were blaming the Indonesians; the few foreign companies still operating in Indonesia blamed corruption in the government, and the Indonesians claimed they were being exploited. The country was on the verge of bankruptcy—though before the war she had exported important percentages of the world's rubber, tin, petroleum, pepper, cloves, nutmeg, quinine, coffee, tea, palm oil and copra. Whole sections of the country were terrorized by bandits, and villages barricaded themselves behind bamboo enclosures....

The Darul Islam, a group of politico-religious dissidents that demands an Islamic state with Koranic law, has terrorized whole sections of the country, burning, looting, murdering and torturing, robbing trains and buses, using every means to undermine the government. Insurrections and banditry have kept the country in a state of chaos right up to the present....

This was not the Indonesia we had come to see, but it is the political and economic climate in which we traveled for thirteen months. The Indonesia we had come to see was that of the old Dutchman in Callao.

'Paradise on earth.' That's a potent lure....

It had been a fast trip from Calcutta to Djakarta, seven days with no ports of call. We had hoped for a more leisurely voyage with time to continue our study of Bahasa Indonesia, the new national language, a sophisticated form of Malay, for centuries the *lingua franca* of the coastal regions of Southeast Asia. Instead we spent the days servicing Tortuga, which, as she was both slow and amphibious, we had named for the Spanish word for turtle.

Helen and I went on deck as the *Drente* entered the breakwater. The water was an oil-green, in the still air smooth as syrup. Near shore a wreck, a casualty of World War II, lay bottom up; I thought of a great dead whale. Some small boys were fishing from its sides and their bamboo poles made yellow streaks against the rust.

On the main deck Tortuga was ready to be unloaded. Dinah, our German shepherd, who had the run of the ship, was lying in the shade of Tortuga's overhanging bow. She had watched us packing, and the last few days had stayed close to the Jeep. She had spent most of her eleven years in one Jeep or another, and she seemed to know that something was about to happen. She watched a sea gull circle low over the ship. She sniffed toward land.

Beyond the long low warehouses, salt-white in the sun, and the row of tall cranes leaning over the water like praying mantises, we could see the harbor master's tower at the old port where many years ago sailing ships from all over the world set up shop on their decks and traded for pepper and nutmeg and clove. Only it was Batavia then, a trading post that was also a fort, founded by the Dutch in the early 1600s. . . .

Today the old harbor is silted in, the fort is gone, and where noble square-riggers were careened in what Captain Cook wrote was the best marine yard in the world, small sailing *prahus* cluster in an abstraction of masts and ropes and gaudy hulls. Where the silks of Persia met the nutmeg of Ambon, the porcelain of China the pepper of Sumatra, and the glass of Venice the sandalwood of Timor, now fish in a hundred colors and varieties are hawked from bamboo stalls. Now bales of rubber, ingots of tin, boxes of tea and bags of coffee are traded for cement and steel and automobiles, and Batavia is Djakarta.

The *Drente* was alongside the concrete pier; her hawsers and rat guards were secured and the gangway lowered. It would be some time before Tortuga could be swung over the side; nothing moved fast in the sultry heat, least of all the longshoremen. With the easy informality that makes freighter travel so enjoyable the captain invited us into the wardroom for a farewell drink.

'A glass of Genever before you leave us?' he asked, pronouncing the name of the Dutch gin so it sounded like 'you never' and then pouring it as if one never refused, even at so early an hour. Apparently one custom had survived.

The captain and all the officers were Dutch, old Indonesian hands, and all the way from India they had filled us with tales of how it was in the old days. Things were not the same since independence, they reminisced sadly; not the same at all. They spoke with an air of wonderment as if they could not understand what had happened to their paradise on earth.

'Why, before the war,' one of them told us, 'when we made port in Batavia, we'd all go to the Des Indes for a *rijsttafel.* We'd have some beer, Heineken's, as good as we make in Holland, and then the food would come.' His eyes took on a misty look. 'Can you imagine,' he continued, 'a whole line of boys, each with a different dish, twenty or more, boiled, fried, steamed, baked, or curried, meats and vegetables, and a great big mound of rice and hot pepper sauces that sharpen your tongue and make you want more. And there was always a boy there to fill your glass with more cold beer.' He leaned back and cradled his stomach. 'That was food.

'Now we don't go ashore much when we're in port. "Incidents," I think your papers call them, sometimes happen, even in downtown Djakarta. A while ago a KLM pilot was killed, shot while he was driving along the street. Of course for you, not being Dutch, it may be different.'

We were to learn that not being Dutch was no insurance when a white skin is your only identification....

The neon lights that outline its modern façade were just blazing into life when we arrived at Djakarta's premier hotel. But despite our reservations there were no rooms, and the clerk was noncommittal as to when we could savor the delights of the famous Hotel Des Indes, Pearl of the Orient, now renamed the Duta Indonesia.

As calmly as if such things happened every day, which as we learned later they did, Mohasson[1] telephoned each of the hotels on his recommended list and many that were not recommended. All were full. Each ministry of the government has its guest house, but they were full too. Gratefully we accepted Mohasson's invitation to spend the night with him. 'That is, if you don't mind sharing the room with my two children,' he added hesitantly.

[1]The Schreiders' liaison at the Ministry of Information.

Mohasson lived in the respectable old former Dutch residential district, an area of stuccoed, tile-roofed houses with leaded windows overlooking canals. He had the top floor of a small but neat two-story dwelling. Mrs Mohasson was expecting us; dinner, she said, was almost ready. . . .

Later, sitting in rattan chairs, surrounded by planters of orchids and wall hangings of handloomed cloth from many parts of the Archipelago, we listened for a while to Radio Republic Indonesia's nightly broadcast of popular American music, a program that a few months later was discontinued as being detrimental to the development of Indonesia's national identity. Briefly we discussed our travel plans, and in his soft, emotionless voice Mohasson expressed doubt, despite our long correspondence with the government, that we would be able to leave Djakarta as planned at the end of ten days. . . .

The next morning we were introduced to Mrs Saleh, Public Relations Officer for the Ministry of Information. All foreign journalists had to obtain clearance from the Ministry of Information, and it was the responsibility of Mrs Saleh to hear their requests. A slim, angular woman with her hair in a stern bun at the back, Mrs Saleh carried this responsibility with an air of resolute and perpetual remorse which seemed to say before she was even asked, 'I'm sorry, there is nothing I can do,' an attitude that was reflected in the dejected expressions of several correspondents who were waiting in her office.

Mrs Saleh's face was a study in gloom as she listened to our plans to visit the major areas of her country: Sumatra, Java, Borneo, Celebes, the Spice Islands, and the Lesser Sunda Islands. . . .

'Well,' she replied, handing us a number of multipage forms to be filled out in quadruplicate, 'why not come back next week? I may have some news for you then.'

Her gloom was catching. I made a feeble request that since housing was such a problem perhaps she could expedite matters.

'Now Mr Schreider, we must be patient. There is much to do.'

It seemed that not quite all had been arranged. We would need a permit to take photographs, a permit to leave the city, and, since Indonesia was still under martial law, a clearance from the Ministry of Information countersigned by the military.

'And don't forget to register with the police,' she concluded. . . .

Our most immediate problem—after registering with the police—was finding a place to stay. Another round of the hotels produced nothing, not even a promise, and though Mohasson had graciously extended his invitation for as long as we were in Djakarta, we knew that we were overpopulating his small home.

At the new glass-fronted American Embassy—later it was stoned by a Communist-led mob of students—we learned that our problem was not at all uncommon. Many a stranded tourist with reservations at some hotel or another had slept on the couches in the foyer, and the guest rooms of State Department personnel were rarely vacant. Bob Lindquist, then Second Secretary of the Embassy, was the first to come to our rescue.

'My guest room is empty until Friday,' he offered. 'After that we'll see what we can do. By the way, how long do you expect to be in Djakarta?'

'Not over ten days,' I replied.

Bob grinned. 'My, you have just arrived, haven't you?'. . . .

The following Monday Mrs Saleh's expression was even more resolute than before. Our permits were still not forthcoming, and her only advice was that we must be patient.

'I really doubt that we can do anything until after Independence Day,' she explained.

'But Independence Day is three weeks away,' Helen and I chorused. . . .

We had only until December to reach Timor; we could not afford to waste more time in Djakarta. But when we explained this to Mrs Saleh, she merely shrugged. We decided to see what we could do by ourselves toward getting those permits. Exhibiting as much patience as we could muster—to do otherwise is in poor taste in Indonesia—we made the rounds of the various ministries whose stamp of approval we needed. We sat at the low tables that are fixtures in every office, sipped the sweet lukewarm tea, made small talk about the weather or the traffic, alluded to the subject of our visit, and finally got around to the details. Without exception the officials were courteous, sympathetic and full of helpful suggestions, and when Independence Day rolled around we were—That's right—we were still in Djakarta. . . .

That evening we discussed the pressing matter of permits with our latest host, Colonel Ray Cole, United States Army Attaché. A big rugged man who speaks little but smiles easily, Colonel Cole was of the opinion that he was a guest in Indonesia, and as such he should do things the Indonesian way, a philosophy that made him very popular with the Indonesian officers with whom he associated.

When we had brought him up to date he smiled, and in his casual way, said, 'Well now, that shouldn't be too difficult. Let's see what we can do at a party or two.'

With that our 'business' hours changed from daytime to night as we accompanied Colonel Cole on a round of official receptions. Each night we met a different dignitary, and by the end of a week we had all of our permits plus a dozen introductory letters. But most important we had a letter from Colonel Soenarjo of Military Intelligence, next to the Ministry of Defense the most powerful agency in the country, requesting all authorities to assist us.

As we picked up the last permit, a smiling Indonesian official said, 'What a pity we did not meet five weeks ago. In Indonesia you must always do things through friends.'...

Away from the suffocation of Djakarta's delicately balanced protocol and red tape, its tensions and superficial gaiety, its restrictions and registrations, we felt a new lightness and freedom....

We came upon Pelabuhan Ratu suddenly, almost startlingly. The road turned, the forest opened, and there it was: one paved street dead-ending at the sea, still marked with a starting and finish line for the footraces held on Independence Day, two rows of frame houses with blistered paint, windows framed with plants in margarine tins, a jumble of kampongs to the sides, the whole flanked with flooded rice fields that lay like gleaming panes of scarlet glass in the afternoon light. The sea was a Monet lake stippled by the sun, the sky a canvas of bold brushstrokes in crimson, the beach peopled by Gauguin. Nets were stretched to dry where an armada of high-peaked prahus was drawn up on the sand to await the tide. It was one of those scenes that is too beautiful, too like a travel poster. I reminded myself that those puddles between the houses that picked up every bit of color from the sky were open sewers, pools of fecundity for mosquitoes.

We looked for the inn, followed by a growing parade of wobbling bicycles, pairs of youths hand in hand, and young children, the girls in loose cotton print dresses, the boys in nothing, everyone kicking at the yapping bony dogs whose sleep on the warm black asphalt we had disturbed.

We were the only guests at the blue, one-story frame inn. It was clean, with little cells for rooms, and there was a large room open to the air from the sea, with a long bare table for dining. There was no food, but the innkeeper, a young man in his twenties, sent for some from the Chinese shop down the street: *nasi goreng*, rice fried in coconut oil and flecked with bits of dried fish, red chili and hard-boiled egg. And there was coffee, thick and sweet with the rich flavor of the fire-roasted bean....

Beyond Bogor [formerly Buitenzorg] the road climbed steeply into the mountains, part of the chain of over four hundred volcanoes that stretches the length of the Archipelago. Motorized traffic was light as we drove on, but hundreds of bicycles thronged the roads. Women pedaled leisurely, somehow unencumbered by their tight kains, easily balancing on their heads heavy baskets of mango or papaya or dried fish; soldiers, Sten guns slung from their shoulders, rolled in squads; children, their faces scrubbed, their hair combed, their shirts and shorts neat, hurried to the schools that almost every village now has....

The air grew perceptibly cooler as we wound higher into the mountains through tea estates and rice fields, casuarinas and giant ferns, toward the resorts of 4,800-foot Puntjak Pass, a weekend target for enervated Djakartans, Indonesian and European alike. Gaudy bungalows in primary colors, interspersed with Swiss-style chalets and restaurants, clung to the steep hillsides, while here and there a spot of turquoise from a swimming pool broke the smooth green of a golf course. At the top of the pass we dug sweaters from Tortuga's cabinets. After the heavy heat of the lowlands, the cold wet mist that rolled in white puffs over the vast panorama of green seemed to wash away the haze on our spirits. We stood there a long time—until a military convoy bound for Bandung stopped and advised us to go on. This was Darul

Islam country, they warned, not a pleasant place to be after dark.

Bandung, with its hermaphrodite ferro-concrete architecture, by Indonesian standards the country's handsomest city, lies on a broad, undulating 3,000-foot-high plateau. Flooded rice fields stretch everywhere, intruding on the city limits where new buildings cast their shadow across heaving buffalo plowing hock-deep in mud that is as dark as melted chocolate. But away from town, rice follows its ageless pattern, spreading across the valleys a patchwork of multihued green, and ascending the hills in stepped terraces. Wherever there is water, rice is a year-round occupation, and in places we saw the whole cycle at a glance: the women transplanting the spring-green rice seedlings; in adjacent fields the maturing crop, that rich green that only rice can achieve; and the dry fields, golden and heavy with bursting kernels, clamorous with the bright confetti of rainbow-clothed harvesting girls. Deftly they palm their *ani-ani*, the crescent-shaped rice knives, concealing them so as not to offend the Rice Goddess, whose blessing is needed for the rice to germinate. Swiftly they cut the stalks; laughing they carry them to the village cooperative. Harvest is a happy time, for rice is life to Java....

Bandung lies in the heart of Java's most troubled area; the Darul Islam were making raids right to the suburbs of the city. Much of the road from Bandung to Jogjakarta in central Java was classified as insecure; villages were barricaded and the few trucks and buses that risked the drive were often burned. On the advice of the military we detoured north toward the Java Sea.

Near the high hill that dominates Bandung's northern limits there was ample evidence of the depredations of the Darul Islam: burned army barracks, shattered windows, yellowed masonry pocked with bullet holes. Armored cars waited beside the twisting road, shaded under dun-dry canopies of palm fronds. Like jack-in-the-boxes their turrets popped open as Tortuga passed, and helmeted soldiers of Indonesia's crack Siliwangi division answered our waves with thumbs up and cries of '*Bagus, tuan*' ['Good, boss']. Being shot at once was enough and near the military posts we observed strictly the speed limit—six miles per hour—and then hurried on, trying to make the safety of the coast before dusk....

Java's north coast, with the shallow sea fading from brown to green to hazy blue, was the scene of the first Japanese landings on the island. Rusted hulks of landing craft still littered the shore, but in the coastal town of Tjirebon [Ceribon] the anti-litter campaign was in full swing, part of the Army's attempt to teach merdeka [freedom]-conscious Indonesians some civic responsibility. The streets were patrolled by rifle-toting soldiers, all traffic was stopped for an hour while the employees of the British American Tobacco Company, at full pay by the company, of course, swept the streets clean. In other cities in Indonesia the same thing was tried, and in Djakarta when sirens sounded all cars on the street were required to stop, and their passengers and drivers to get out and clean the area around them. The system was changed somewhat when a diplomat objected to being straw-bossed by a private with a Sten gun.

At the Grand Hotel Tjirebon, an immaculate hotel on the main street, we were relaxing over a bottle of still-excellent Heineken's when a group of teen-agers approached. With their white shirts and polished black shoes, their tight trousers low on their lean hips and supported by nothing more obvious than faith, they were as alike as club members—which they were. Would we address their youth association that evening? They had been mightily impressed by four youngsters sent to Indonesia by the American Field Service on an exchange visit of high school students. All were anxious to practice their English, hoping for a chance at the next exchange.

But the assembly of more than five persons without special clearance was still prohibited by the military; permission was refused, and that evening we all sat on the lawn in front of the hotel. More youngsters, including a few girls in crisp, Western-cut batik dresses, their long hair tufted in glossy black pony tails, drifted in until there were a score or more, and Helen and I expected any moment to see a squad of police come to disperse us. But though soldiers passed they looked the other way, and we talked until curfew, answering questions about public schools in America, the cost of living for students, about Little Rock, Hollywood and Elvis....

Sometimes on those side trips we stayed in *pasangrahans*, government rest houses for official travelers built when Indonesia was a colony but today often roosting places for bats. Sometimes we stayed in *losmens*, small inns where we could generally follow our noses to the 'sanitary' facilities and slept in airless cubicles under tattered gray mosquito netting. Once, in a town in west Java, the only room available was an eight-foot-square shaft with a tiny barred opening near the ceiling. Stained whitewash peeled from the barren walls, and phalanxes of cockroaches advanced on us when we entered.

Helen took one look and declared like Papa Bear, 'Somebody's been sleeping in my bed.' We would have been happier in Tortuga, but this was Darul Islam country and there was nowhere else to stay. Exhausted by a hard drive, energy sapped by the humid heat, I told Helen it was her imagination, to go to sleep. But she doggedly examined the bed by candlelight and I called the slovenly clerk away from his card game outside our room.

'This bed has been slept in,' she told him, indignantly.

A surprised expression sheathed his face. '*Tetapi, tuan, satu orang sadja* (But, sir, only one man).'

'Well, that's one man too many,' Helen retorted, and the clerk grudgingly changed the one sheet, muttering all the while. For the rest of the night we tossed, fanning the still air to drive away the mosquitoes attacking from above and slapping the bedbugs boring from below.

But more often, once we were in the secure areas of central Java where the Darul Islam had not penetrated, we preferred Tortuga's facilities to anything the losmens had to offer. We would duck into a stand of teak, a clump of palms, or higher in the mountains a grove of casuarinas overhung with Spanish moss and alive with all the squeals and whistles so abundant in a tropical forest.

Satisfied that we were unobserved, we would select tinned food from Tortuga's cabinets, uncover the sink and alcohol stove built into the dash, pump our teakettle full of water piped from the tank in the bow, and while Dinah explored or relished her supper of kibble and tinned horsemeat, we relaxed and read until it was dark. We would extend our foam-padded bunks from behind the seats, and protected by the screens over the windows, we would sleep easily until dawn.

But sometimes, soon after we thought we were unseen, Dinah would growl. A face would peer from the bamboo, and a child, drawn by that unexplainable aura that exudes from anything or anyone strange to the countryside, would go running to his village, crying, '*Orang putih, orang putih, kapal dharat* (White people, land ship),' and the elders would come to investigate, no matter how far.

Speaking through the schoolchildren—the elders knew little Bahasa Indonesia—we would explain our presence and our nationality. 'No, we are not Dutch. We are Americans; yes, our kapal dharat goes on the sea and the land.'

Astounded they would walk around the Jeep; smiles would replace scowls, and the thumbs would go up. '*Bagus, tuan, selamat datang*' [welcome], and they often invited us to their village.

Perhaps there would be a Moslem circumcision ceremony in the village that night, or a marriage with the groom and bride dressed as prince and princess of the royal families that are still revered though deprived of temporal power in the new republic. Or perhaps there was a *selamatan*, the ceremonial feast of rice cakes in all shapes and colors, of syrupy sweets, and curried buffalo meat from an animal sacrificed in the traditional one-stroke decapitation that is to appease the spirits on the opening of any festivity, for even devout Indonesian Moslems still carry in their hearts the beliefs of their animist forefathers.

But even if there was nothing special we were usually invited for tea. Once we walked along the narrow dikes between flooded rice fields, joining the good-natured laughter when we slipped off the edge into knee-deep mud, past flocks of brown dappled ducks trained to stay near the little white flags in their midst, to a palm-hidden village of rectangular cane-and-thatch houses. The mosque, open-sided with a two-tiered roof, was the most prominent structure, but the huge wooden drum for calling the people to prayer was silent and only one man knelt on the bare wooden floor inside.

The village chief, a diminutive ancient whose bones jutted beneath parchment skin, waved us into chairs in the hut's front room. His wife, equally ancient, but arrow-straight from a lifetime of carrying heavy loads on her head, brought tea, each glass capped with an embossed aluminum cover to keep out the flies that droned around us.

As we waited for the invitation to drink, our eyes wandered

from the inevitable picture of President Sukarno on the thatched wall to the rolled sleeping mats in the corner, the string of dull red chili peppers hanging from the rafters, the charcoal brazier glowing through the door to the cooking room where the number two wife, a smear of green paste on her brow to cure headache, prepared the supper of sago, rice, fried bananas and dried fish.

The door and the unglazed windows were dark with children, the young ones healthily clad in nothing, all eager for a look at the strangers, the orang putih. Beyond them an old man, a neat, scanty turban of blue and brown batik on his head, his age-corrugated chest bare and his sarong pulled to his knees, hunkered on the ground and fondled his fighting rooster, probably ruminating on its former conquests, for cock-fighting and its accompanying gambling, once so popular as to be almost a vice, has been banned on Java, and is permitted only on holidays on the neighboring island of Bali.

Finally, when the tea was cold, the old chief said, '*Selakan*'; we drank, and our glasses were immediately refilled. We ate the rice cakes he offered, and, anxious to show him we enjoyed it, we finished our tea. Again the glasses were refilled, despite our protests. These were not demitasses. They were large tumblers, and I could see the distress in Helen's eyes saying plainly, 'I just can't drink any more.' It was months and gallons of tea later before we learned that a sip and a nibble are all that etiquette requires. . . .

It was one of those heat-pregnant evenings in Djakarta when the afternoon rain has stopped and the night rain has not yet begun, when everything seems poised, as if waiting for that first stab of lightning, that first clap of thunder, those daggers of rain that in a few moments flood the streets and sluice off into the old Dutch-built canals. Dinah sniffed the air in apprehension, a drop of nervous moisture hanging suspensefully on her long nose. I watched the drop grow, then splatter to the tile floor to be followed by another. She was a good weather prophet—her one fear was thunder. That and being left behind when we started on another journey.

The electric fan, an archaic model, swung idly back and forth, back and forth, an almost hypnotic movement, rustling the muslin-thin drapes and stirring my disordered pile of notes and travel permits on the table. The fan slowed, then stopped; the lights, a pale yellow at their brightest, flickered and faded to orange to red to darkness as the electricity was cut off. I lit a candle and hurried to the bathroom. It was all right; the mandi, the large cubical tub, was filled with water. Here in the Hotel Duta Indonesia, formerly the Des Indes, Pearl of the Orient, there would be no more water or electricity until morning.

Across the room was the sleeping area, a screened-in cubicle like a cell, with kapok pads over wooden slats and two iron bedsteads painted a sickroom cream. On one, Helen was tossing in her sleep, her form appealingly outlined by the sheet, appearing almost ephemeral in the guttering candlelight, elusive behind the glowing copper screen. I wished it were not so hot.

The drapes were still now, and I stood by the window in my shorts, my body damp with perspiration. How futile to dry after a bath; the exertion of moving the towel over your body made it wet all over again. The wet smell of bougainvillea, canarium, palms and frangipangi drifted from the ground below. Frangipangi, to Indonesians the flower of the dead. I rubbed my stomach. It was still sore from having a Sten gun jammed into it that morning when I tried to get at the soldier who clubbed Helen with a rifle butt. Helen's concept of justice had brought us both very close to death at that moment.

The canal that bisected the street in front of the hotel was several hundred yards away, too far to add its questionable fragrance to the perfume of night outside my window, but I could see the tiny bobbing lights of the *tukang sate* [saté vendor] moving along the street. Some of those miniature kabobs, the charcoal-broiled tidbits of chicken or beef skewered on a bamboo sliver and dipped in peanut sauce, would taste good. But I was too lethargic to call the room boy to send for some. He probably would not have come anyway. I knew he was sleeping at the bottom of the stairs, but this was the age of merdeka, freedom. The cries of the tukang sate, the staccato squeaks, *e*, *ee*, *eee*, faded into the night and were gone.

A shot echoed from the direction of the banking district. Someone must have missed the sign; the area was closed by the military after six p.m., but the warning notice was small, the barricade only a coil of barbed wire to one side of the street and the streetlights were dim. I hoped whoever it was had not been killed. The last one had, a European it was, a newcomer who knew not a word of Bahasa Indonesia, who could not have read the sign even if he had seen it. The silence in the streets now was a tension that could be plucked, like a guitar string. The Indonesians had learned to live with that tension and so had we, but we did not cover up so well. . . .

That face, I wish I could forget that face—round, pock-marked, little eyes hiding under heavy overhanging brows, short neck, long arms, stocky body, that green uniform. And the Sten gun, God, how loud that click sounded when he cocked it, like a breech block rammed home on a field-piece. . . . I shouldn't have rushed him. But what could I do, Helen crying, holding her breast, I'm so grateful the doctor says she's all right, that the bruise will disappear in a few days. That was as close as I ever want to come to it. Bastard—he was shaking so much he could have pulled the trigger by accident. And then all he would have said was we were resisting arrest and nothing more would have been done. Probably would have given him a medal. Resisting arrest, what a laugh. What right had he to get into the Jeep? And then to slug Helen when she puts out her hand to stop him. Just showing off, a corporal showing three privates how to treat the white foreigners. He didn't even know the name of his commanding officer; there it was right on our pass. He looks at it three times, lets us through the gate three times, in and out, to the customs office at the dock, to the shipping office, back to customs. And when we have only twenty minutes to get to the main office to pay for our tickets, fourth time through the gate, he says our pass is no good, says we have to take him to headquarters and starts to climb in. When Helen said no, that rifle butt came back so fast I didn't even see it. A woman saying no to him, a white woman at that. I guess he lost too much face in front of his privates and he slugged her. And when we take him to headquarters he keeps that gun pointed at us all the time. And the lieutenant won't even let us call the Embassy. Holds us there an hour. We'd still be there if we hadn't been able to call Hari. It's lucky that he's a major and could vouch for us. . . . I wish I could sleep. I'm shaking all over again. I have to forget this.

30
Maslyn Williams Seeks the Truth in Sukarno's Indonesia, 1965

Java was on the brink of catastrophe in the mid-1960s when Australian author Maslyn Williams made his five journeys from Jakarta. Soon a coup d'état *would sweep Sukarno from power and precipitate the massacre of hundreds of thousands of the island's people, most of them affiliated with its large and popular Communist movement. In the passages below, Williams captures the surrealistic hopelessness of the pre-*coup *days and provides a glimpse of Sukarno's charismatic attraction to the people of Java.*

FASTEN seat belts.

The aircraft turns in over the coast, slants across the ships in the outer harbor, and drops down over the city. Roof-tile red and the brilliant living green of tropical grass, and palm trees crowding the edge of the airport. It is like coming in low over a village. Boxlike huts with palm-mat walls and dirt floors huddled together at the end of the landing field, leaning one against the other for support. Branchless papaya trees with clusters of green and yellow fruit hanging at the top under a few sparse leaves. Bananas. Chickens scrabbling in the dust and brown dogs with lean ribs foraging without enthusiasm. Naked babies sitting on the bare ground scratching patterns with pieces of stick. Alongside the runway, as we roar along with flaps standing upright and quivering in the slipstream, two men sickle the long grass and hang it upon sticks. They have paid some official for the right to do this and will sell the grass to men in the villages who have oxen or ponies to feed.

Near some tin sheds there is a nest of slender antiaircraft guns, pointing like pencils into the sky. Each gun has a low wall of sandbags round it. Nissen [a companion] cranes forward to watch through the windows as we touch down but makes no comment as the guns flash past. But we can hear the Englishman behind us 'They come from Czechoslovakia, not paid for, of course, and they don't work. No spare parts. It's the same with everything in this country.' There is no malice in his voice, no ill will, not even condescension. Just complacency and the certainty that everything here is inferior and of no consequence.

Outside is a hot, gusty wind full of dry dust. It tugs at the row of flags flying on high staffs in front of the airport. There are more than fifty of these flags, half of them Indonesian (red and white) and the other half Russian. Nissen says that [Anastas] Mikoyan has come from Moscow to visit the president. It is, he says, a matter of armaments. Walking across the tarmac we pass aircraft belonging to the Indonesian air force. Some of them come from Russia, others from the United States, and some from Britain.

Young men in uniform are at every gateway and entrance and walk about in the foyer. All of them carry automatic rifles and look serious, but there is no trouble. One wonders why this charade of armed watchfulness should be necessary and suspects that it is a matter of keeping the young men employed. But the girls at the currency control tables are friendly and smiling, and the customs clerk marks my bags without asking me to open them. Outside in the foyer a gesticulating taxi driver aims his way at me through the crowd.

'You go where?'

'Kebajoran [Kebayoran].'

'Five thousand rupiah.'

'Don't be silly.'

'How much?'

'One thousand.'

'Okay.'

He takes up my bags and pushes back through the crowd. His car is very old and painted a vivid green with plain house paint. None of the windows can be adjusted. One of the back doors is tied shut with rope. He is clearly not a licensed driver, and this is not a registered cab. But people must eat. This is the first rule of the Republic. The West can afford to insist on standards of administrative order and efficiency, can even make a fetish of it, but here the first need is to survive, and after that to try to catch up with a technological civilization which was developed while generation after generation of Indonesians was held to a level of a medieval peasantry.

It is easy for us of the West to criticize. It is also, of course, just as easy to make too many excuses. And it is going to be difficult to recognize the truth that I am seeking. . . .

Six lanes of traffic speed along the highway between Kebajoran, where the middle classes live, and Jakarta, where they work. I sat beside the driver of a spacious, light-gray Mercedes sent by the Foreign Office. Two other passengers were in the back seat, one of them a young diplomat recently returned from Egypt, sent to fetch me for an interview with his chief, the other a Traffic Department official who watched out of the window with a look of studied concentration.

We were hemmed in tightly, going with the fast flow of early morning traffic like a stick in a swift stream. On one side a bare-framed vintage jeep kept pace, swaying alternately toward and away from us as though dancing a *pas de deux*, while the driver talked over his shoulder to passengers bulging out of the back; on the other side a Cadillac, all shine and sheen of newness, smoothly carrying two Asian men, expressionless with the impassive, protected look of those who have no worries about money, sex, or heaven; and in front of us a rattling, sagging, asthmatical bus breathing black smoke like a story-book dragon.

The traffic is mostly raffish, unruly, and without discipline—cannibalized vehicles of every make and age mixed in with horse-drawn drays and wagons, bicycles, bemos [motorized pedicab], and a stream of *betjaks* (the tricycle rickshaw of Indonesia). They all go with no regard for rules, pass each other on either side indiscriminately, compete for loopholes and openings in the flow, cut in, swerve, swoop, and stop suddenly; it seems disorderly, irresponsible, and dangerous although there are few collisions and seemingly little ill temper.

Every few minutes we pass a breakdown being rushed to

the side, opened up for examination, or abandoned. It is not a matter of mechanical ineptitude but of economics; the shortage of foreign funds. Parts wear out and cars break down like worn-out horses. Nothing can be done. Riding into town, one begins to wonder if, after all, the critics are not justified, wagging their heads and saying that the country is running down, cannot go on much longer, must collapse economically. What then? Civil war and victory for the Communists.

'In many countries,' said the man from the Traffic Department, 'they use television to observe and control the traffic. I have studied this subject. It is very interesting.' He looked out of the window again.

A young man riding a motorcycle swung past us, weaving a mazy way through the traffic. He sat arrogantly upright, hands in the pockets of a black leather jacket, guiding the machine by the swaying of his shoulders and the pressure of his thighs. It seemed suicidal.

The driver beside me scowled and muttered, 'Cross Boy' (the local synonym for far-out fellows), adding that he should be in Borneo with the volunteers, exercising his recklessness on the British. The man from the Foreign Office laughed; but the Traffic man agreed with the driver, adding that the youth ought to be arrested and dealt with. 'In any Western country where the traffic is properly organized, that boy would be in prison.' But he spoke without conviction, as though the notion of putting anyone in prison was fundamentally foreign and repugnant.

Further on the traffic stopped, coagulated into a question mark. We speculated: a breakdown, an accident, a police search? There were rifle shots, the sound of mortar shells exploding, of whistles being blown. Word came back, passed from car to car, that it was an army exercise.

We sat, hemmed in for half an hour with the car windows shut to close out the smoke coming from the bus shuddering in front of us. We took off our jackets and fanned ourselves with papers. Apart from the discomfort nobody seemed concerned that a military maneuver should be timed to coincide with peak hour traffic on the only city expressway.

Soldiers suddenly appeared among us, using the cars as cover: young men in jungle uniforms sprouting stiff sprigs of camouflage, twigs and leaves in their helmets or poked into their clothing to break the shape and form of figure. Their faces were daubed with dirt, and all stared ahead with fixed and seemingly deadly intent.

Most of the people sitting in cars looked bored although in no hurry, but our man from the Foreign Office rolled down the window and spoke to a young soldier crouching alongside, asking him about the exercise. The soldier answered over his shoulder, urgently, making the situation seem almost real, 'We are attacking the telephone exchange.'

Whistles blew and the young soldier scuttled on, hunched over his rifle, threading a swift, twisting way among the stationary cars until he came, with many others, to the edge of the expressway. Here they threw themselves down on the grass, pretending to enfilade a block of square, white buildings a hundred yards away. Between them and their objective imitation shell bursts threw up little fountains of smoke and earth. The man from the Foreign Office shook his head. 'They are crazy. This telephone exchange is a very inefficient installation. Whoever holds it will be handicapped and lose the battle. Better to destroy it than to capture it. Better still, leave it in the hands of the enemy.'...

Later in the day I was taken to the Blueprint Hall, a beautifully designed modern building of six stories standing on the spot where President Sukarno proclaimed the independence of Indonesia on August 17, 1945. Each story is seventeen meters (approximately fifty-six feet) high. A column in front of the building is shaped like a crowbar, signifying that independence must be followed by hard work and national development, and on top of the column is a symbolic shaft of lightning, implying that the declaration of Indonesian independence struck the world with electrifying force.

Inside this building is the vision of the Indonesia that is to be, all set out on the six floors—hundreds of fine, well-made models of hydroelectric water supply and irrigation schemes, mines and steelworks, schools, universities, hospitals, laboratories, railway and road systems, docks, harbors, shipyards, bridges, factories, tourist centers with huge hotels; telecommunication systems, television stations, air-radio and navigational aids; giant tanks, guns, rockets, huge flights of warplanes, battleships, nuclear reactors; and as a background to all this, maps, charts, graphs, photographs, lists of statistics, slogans, and excerpts from the President's speeches.

It is an astonishing sight, this huge, beautiful building filled with something that doesn't exist. As I stood there wondering where to start, a line from a poem I had long forgotten came back into my mind. 'Tread softly because you tread on my dreams.'

Usually the building is busy and filled with the murmur and continuous movement of people being guided about in groups, or parties of school children with their teachers, but it was closed now for a period of rearrangement, and empty of people, so that I stood alone in the first great gallery waiting for a guide (the director was anxious that I should be shown everything and have it clearly explained in English).

He came with four girls, shy but excited to have a foreign writer to escort. The director said, 'They can all speak English,' and the girls giggled and twittered and said that their English was poor but that among them they would manage. I felt the better for having them there, believing that I could manage girls so young and bend them to my own inclination in the matter of what to see and what not to waste time over, but I was much mistaken for as soon as I started off to look at something in the middle of the gallery they all cried out in surprise and alarm and jostled me back to the starting point like shepherd dogs retrieving a recalcitrant old ram. I made gentle attempts to break away, smiling at them and pretending not to understand what they were at, but they would not meet me on this and steered me back to start the tour properly, from the beginning.

It took three hours in all and I was exhausted and for the last hour took nothing in, although they continued to be gaily relentless and inflexible and never for a second let me rest or gloss over any part of the display. Much of it was above their heads and I soon ceased to ask questions, even out of politeness, because they had no considered answers, only statistics and official explanations learned by rote, recited guidebook fashion and interlarded with caption-quotes. 'To build a Just and Prosperous Society'. 'Listen to the Message of the People's Suffering'. 'Our Struggle'. 'Our Plan'. 'Our Great Leader'. In a while the words became irritating coming from the lips of these pretty children, and I began to think of other things.

I stood at last in front of models of a spaceship and an orbiting satellite set against a huge photo-mural of the night sky with a caption reading, 'The People of Indonesia Must Be Space Conscious'. Out in the fields the farmers plough with wooden ploughs and oxen, and the women reap the rice ears by hand. And in Jakarta there are no water closets or shower baths or footpaths to the streets, and one cannot buy toilet paper. So 'The People of Indonesia Must Be Space Conscious!'

We of the West have more subtle idiocies. Such elemental self-deception as this is too much for our kind of mind to encompass. I could understand these girls being in some juvenile, unthinking way inspired by all this and proud to belong to a newly emerging nation vital with humanitarian ideas and ambitions, but that national leaders and trained experts should solemnly sponsor this make-believe, this shadow-play, seemed to me an exaggerated way of escaping from reality; a form of dreaming that might end in disaster. But the same was said of the Russians fifty years ago!

Before I left the building the girls gave me coffee and cakes, served on a balcony overlooking the spot where the president had proclaimed the Republic, and they made me promise to send them my book when it was written, but first to come back in a month, when the whole exhibition would have been changed and brought up-to-date with new models of other projects and objectives....

On the third morning of my visit I left Bandung a little before daylight.

The other eight passengers in the little bus seemed surprised to find a foreigner traveling with them, although they responded promptly enough to my 'good morning', and squeezed up willingly to let me in. But once I was settled on the back seat between a youth and a stack of baggage they made no attempt to talk. In any case it was too early for conversation, so we sat mute, each inside himself, knowing that with the long day to fill with each other's company it was best to begin sparingly.

We went westward out of the city, heading across country to Tjirebon [Ceribon], a secondary seaport and naval base; after that we would follow the coastal road to Semarang, capital of Java's central province, where I would stay a few days, the total of this leg of the journey being some 250 miles from Bandung. We traveled by *Suburban*, a small bus which carries a fixed number of people on intercity routes, not stopping to pick up passengers indiscriminately at villages or along the road as other buses do....

When we had been two hours on the road we stopped at an eating place to have breakfast, after which we walked up and down, stretching our legs and getting acquainted, then went back into the bus ready to be civil and communicative. Two women and a child sat in the front seat with the driver; three men sat in the seat behind them: a sugar merchant, a young judge, and a buyer of hides and skins for a leather exporting company. In the back seat beside me, was a student with his young brother and the pile of baggage.

The stopping place had been at the top of a dividing range, and we nosed down now toward the coast, through tea plantations and stands of young rubber trees (thin and spindly, yet with the incipient grace of adolescent girls), then came through teak forests and fields bristling with stiff cassava plants to the outskirts of Tjirebon, where we were stopped at a road block.

A policeman peered in at us and asked for the driver's license, which he took and went away with behind the bus. The sugar merchant said that it was a routine matter, an attempt to prevent overloading and the use of unsafe or unregistered vehicles. He said that the whole motor-transport system of the country was in bad shape since currency restrictions had stopped the importation of spare parts. The number of road-worthy vehicles diminished each week, which meant that the remainder were worked even harder and they, too, soon broke down; in any case, overloaded trucks and buses damaged roads and broke bridges, and for these reasons the traffic check was necessary.

The policeman came back and passed the driver's papers to him through the window and waved us on. The sugar merchant leaned forward and said, 'How much?' Without looking around the driver replied, 'Fifty, it is nothing,' then moved into second gear.

The merchant turned back to me. 'He had the money hidden in the license. It is better to pay a little. If the driver refuses to contribute to the police funds he will be made to take his bus for a technical overhaul, which will take several days and cost a lot of money. His license will be sent to Bandung for checking, and this might take three months. As it is, everybody pays a little, and no time is wasted.' It seemed reasonable until a few miles further on we came to a bridge with broken bearers, sagging so badly that it must soon collapse; and a little further on was a truck so overloaded that the back wheels were beginning to buckle. When we had passed the overloaded truck the skin buyer spoke.

'You see what bribery does—it just encourages the breakdown of all efficiency and order. It was not like this in the Dutch times. The country is falling to pieces. Foreign ships are coming here to Tjirebon or to Sourabaya because the warehouses in Jakarta are full to overflowing and there are not enough trucks running to clear cargo from the wharves.'

The sugar merchant said disgustedly, 'Jakarta, Jakarta, there the trouble lies. Everywhere else in Indonesia people get on with their business, trying to build up the country, but in Jakarta they bring us down with their plotting and political games.'

The student was wriggling, and while the others went on talking he turned to me. 'These old, middle-class men who used to work for the Dutch are all the same, always complaining because they can't make more money at the expense of the people and live like imitation Europeans. They would rather go back to the old times and be comfortable themselves than help us to get justice for everybody. They are worse than foreigners.'

As the student stopped speaking the skin buyer was saying, 'In the old days we had bread for breakfast and maybe a little meat and jam, but now this is too expensive and we can only have rice. At night we could read a decent newspaper and magazines from Holland and America or England and listen to nice music on the radio or visit friends and talk freely about anything; but not now. The newspapers print only government speeches and the radio makes government talk all day long; it is illegal to complain, so why go visiting—there is nothing to talk about.'

The judge spoke, choosing his words slowly. 'It is the nature of men to complain as they get older and to say that things were better when they were young, for this is how we all excuse the failure to achieve our dreams and ambitions. Our fathers were able to criticize the Dutch, but now we have nobody but ourselves to blame for the way things go in our own country, and this makes us bitter and ashamed.'

I asked if he found difficulties in applying the security and emergency regulations aimed at suppressing antigovernment talk and demonstrations, to which he answered that justice is an abstract concept subject to practical pressures. 'In such times as these,' he said, 'a judge must come to terms with his conscience or give up the job. A thief can be sent to prison for a day or for five years or anything in between, depending upon the amount stolen and whether under trivial circumstances or as the result of planned burglary. But a man with no rice field, no money, and many children may steal simply to be sent to prison, where he will be fed and clothed while the government supports his family. It then becomes not a simple matter of crime and punishment but of dealing with a complex social and political problem in which the judge must do what he thinks best for everybody.

'A judge must also be sensitive to political aspects of justice in these days of political emergency. It is a criminal offense to join in subversive talk, to hold illegal meetings, to belong to proscribed political parties, to read various foreign magazines, or to listen to Radio Malaya. But many otherwise law-abiding people do these things, and if somebody reports them they must be arrested and tried. A man who is not involved politically might be sentenced to a single day in prison simply to teach him a lesson, but a man with powerful political enemies who press the case against him may be put in prison for a year for listening to Radio Malaya. If a judge is to remain in his job and do the best he can for the people, he must know when to bend before political winds, remembering that his job is to uphold the laws, whether he agrees with them or not.'

The skin buyer said, 'The government makes laws to cut us off from the rest of the world and from Western culture. In my town we had an English-speaking club which met every week to read English and American magazines and technical publications, to discuss them and learn something, but now we are afraid to meet. We had a Rotary Club, but it was banned because the Communists said it was an instrument of Capitalist propaganda. There is nothing left in the small towns for an educated person, only the shadow plays, which are for peasants.'

The student spoke up so that the men in front could hear him. 'These white Indonesians have the American idea of culture, the culture of *Time Magazine* and Rotary clubs. They forget that this culture [he spoke the word viciously] despises all colored people, including themselves. They are so much in love with this culture that they complain because they cannot get magazines which tell lies about our country and our President. What right do these American barbarians have to tell lies about us, to demand that we stop demonstrating against them when they never cease to slander our leaders? By what right do they claim to be able to tell us what to do, how to run our country, what standards we should set for ourselves? What makes them think that they are so much better than we are? No American will get hurt in a demonstration in Indonesia because we are decent people, but plenty of Americans get beaten up in their own country, especially if they are colored. Let them mind their own business and keep their culture to themselves.'

Flat, shadeless fields stretched far on either side, patterned green, yellow, and black with blocks of high tasseled sugarcane, squares of stubble, and patches of fallow land; and across this toylike landscape went a tiny locomotive pulling a string of rickety trolleys of stacked cane to a long whitewashed mill built on the edge of a stream. Men in the fields with broad-bladed hoes were digging deep trenches, and behind them women came with arms full of sugar stick for planting. For a few moments there was quiet, broken only by the distant whistling of the locomotive.

The judge sighed and said, 'It is not so easy to draw sharp lines between good and bad. Politicians, teachers, journalists, even religious leaders must oversimplify things because they have not the time or capacity to teach people each detail of truth. So they telescope complex ideas into slogans and catchphrases that can be learned quickly and to which people respond like Pavlov's dogs, without even understanding the whole and complete meaning of the ideas that the catchwords represent. The slogans soon become more real than the ideas, and more deadly, because they are more simple and direct; they do not ask people to think or to examine more than one dimension of a situation. Slogans simplify life and make it easy for us to feel that we are doing a reasonable thing even when we know we are doing wrong.

'Students write on a wall "Yankee Go Home". You ask them what this saying means and they tell you, "The Americans are Fascists and Imperialists." You ask again what do these words mean and the answer is, "The United States seeks to rule the whole world with a dollar dictatorship." So it goes on. Each question can be answered with a slogan and the final slogan is always, "I am right, you are wrong." It is the same on all sides. The Americans claim that they seek only to make the world safe for democracy. What kind of democracy do they mean? They are fighting to contain Communism! Is there only one democracy, only one kind of Communism?'

The student leaned forward. 'You are right. We have our own kind of democracy here in Indonesia, "Guided Democracy", and we have our own kind of Communism, "National Communism". We could probably make them work together if other people would stop interfering and making trouble. The Americans try to justify their aggression in Vietnam by claiming to defend Asia against Communism, but we are trying to defend Asia against Imperialism. And there is more justification for our confrontation of Malaysia than there is for the United States interference in Vietnam. We fight against foreign white armies, not against Malayan villagers.'

The skin buyer turned in his seat and spoke directly at the student. 'The Communists have some good ideas and they work hard, but they should not shame the rest of the Indonesian people with their childish demonstrations. If we want the Americans to stay out of our business we should stay out of theirs and not try to tell them how to deal with their Negro troubles and things like that.'

The student, smiling slyly, answered, 'It is the Young Moslems who demonstrated against the United States about racial discrimination, not the Communists. Maybe you get this wrong information from Radio Malaya.' He turned to the judge, joking. 'How much jail will you give him if I report him for listening to Radio Malaya?'...

I went for a walk at dawn, when the city was gray and those who slept in the streets were stirring and saw a young woman in rags, with two children, washing themselves with water that trickled in slimy drips from a waste pipe under a house. The children were teasing each other and laughing, but the mother had no smile on her face. I was ashamed to give her money but called the oldest child, put a few rupiahs into her hand, and went on quickly, remembering a conversation I had had with an Australian journalist in Jakarta soon after arriving in Indonesia.

It had gone along these lines:

Author: How long have you been in Indonesia?

Journalist: Nearly two years.

Author: Have you been able to get around and see much of the country?

Journalist: No, not very much. Most of us are pretty well tied to Jakarta, although I did manage to get to Central Java for a couple of days last year. Just a quick trip there and back.

Author: Well, that would be interesting. It's the center of Javanese history and culture. Did you see anything of the Ramayana or the Wayang plays, or meet any of the painters and dancers or musicians?

Journalist: No, I went down especially to do a piece and get some pictures about the famine and the beggars. There were thousands of them from the dry areas, blocking the streets of the big towns. I think that Ganis [a Foreign Office spokesman] was annoyed about me going.

Author: Have you been anywhere else in Indonesia?

Journalist: No, everything of importance happens in Jakarta. I daren't go away for more than a day or two, at most, in case anything breaks. But I'd like to. It's an interesting country.

Author: But you went four hundred miles to see the beggars and the hungry people.

Journalist: Yes, it was a pretty good story....

The hunger belt in Java is an area south and east of Jogjakarta where the soil is impoverished, the rainfall scanty, and the population excessive for the capacity of the land. There are no industries. In a good year the people grow enough rice to see them through. In a drought year they go very hungry, the old people and little children die, and tens of thousands of others go begging in the streets of Jogjakarta, Surakata, Semarang, and Sourabaya, the big cities of Central and Eastern Java....

The driver said, 'I came down here last full moon and thousands of rats were streaming across the road. If one was killed by a truck the others stopped to eat it. They would stare with their red eyes into the headlights of any truck that came after and refused to move until they had eaten the dead one.' He went on, 'The mayor and the police and army chiefs and men from the offices came and helped the people to put poison in the rat holes, but some of the people ate the poison because they wanted to die.'

The change in the landscape came suddenly. At one moment we were driving through the green Java of wide rice fields, guardian mountains, running water, a man driving ducks to graze, women in line reaping, a man digging with a hoe, another ploughing; the cycle of peasant life, always in motion. Then over a hill, into a dry river bed, and the land was suddenly desolate. The earth was gray, cracked, fractured. Not even a lizard was alive. There were sparse, brittle stalks of stubble, no leaf on any tree, no green weed, dry rice terraces which had the look of antique amphitheatres in Asia Minor.

In a hollow by the side of the road a group of women with buckets and earthen pots clustered around a puddle, the remains of a lake, muddy, thick with scum. These were the stubborn ones who had stayed through the year-long drought. There was a second, smaller pond, five or six yards long and two yards wide. Children were washing in it and drinking the water with their cupped hands. A buffalo with washboard ribs plodded toward it, dragging a small boy by a halter. The animal walked right into the pond and stood in it. It nuzzled a clearing in the scum and began to drink. Then it pissed. The children shouted and thumped it ineffectually with their fists but it took no heed, only switched its tail as if flies were worrying it. The small boy was crying with shame and rage as he kicked at the buffalo with his bare, clay-covered feet.

Coming to a town we passed a procession of thin, dessicated people scuffling along the edges of the broken road, carrying with them such household goods and pieces of furniture as might help them to set up house elsewhere or which might be sold and the money used to keep them alive until rain could bring the rice fields back to life. Some of them carried their things roped to their backs, others had them swung on poles. A man wheeled a table and two chairs on a bicycle, a second man with a bicycle had window frames and a door. A family with a pony cart piled high were heading toward Jogjakarta.

The priest said, 'If they can get to one of the big towns they can pawn their things with the Chinese for rice, and at least there is water in the towns. It is better to suffer together in the streets of the city than alone in an empty village. But if they want to stay here they can sell the furniture and these other things or raise a loan on them at the mayor's office.'

The driver said, 'The mayor buys up their land and whatever else they have to sell and sends them to the towns, where the government has stores of rice for the refugees. If there were enough trucks they could bring the rice here and distribute it around the villages, but the way things are it is easier to take the people to the rice.'

'In Semarang,' said the priest, 'the Communists have organized refugees into bands of beggars and help them to set up road blocks along the main roads so that they can stop the traffic and make travelers pay to get past. If the drivers won't pay they turn over their trucks or cars and strip the wheels off them. But they only do this in places where the local authorities are anti-Communist, so as to create trouble and make the President angry with them for bringing the country to shame.'

There was a transmigration office in the town that we came to, a wooden shed, but few people using it, and these seemed skinny and sad: people making a last gesture of despair rather than of hope—giving up rather than making a fresh start. The man in charge had little enthusiasm for his job and no inclination to talk about it, and to every attempt at conversation said, 'You must ask the mayor, it is his business, I am a clerk.' I could understand the refusal of this man to become involved beyond the mere business of filling in forms, could appreciate his reluctance, a stranger to the district, to let himself begin to have opinions, to feel deeply, to take sides, to try to work out where the answers lie, to allow himself to become, personally, part of the desolation and aridity.

'He is afraid,' said the priest as we drove away. 'He is afraid that the Communists will make trouble for him if he speaks to strangers, so he says nothing, does nothing but what he is paid to do. It is not his business. The big landowners use the famine to take a tighter grip on the peasants, and maybe some of them work in with government officials to encourage the people to sell their land and houses and leave the district. The Communists try to break the grip of the landowners by buying the houses and taking over the peasants' land themselves. They say they are helping the people to get land reform; that is why they organize the begging and the hunger demonstrations. Maybe they are doing good, maybe not. Who knows? Communists cannot be trusted.'

The driver said, not unkindly, 'You are a priest and you must be against the Communists, whatever they do.'

The priest looked at him and asked, 'Are you a Communist?'

The driver eased himself in his seat and spoke, looking straight ahead. 'No, there are enough people who think they have all the answers. I am like the clerk. I do my job and mind my own business. It is best. More people should do the same.'

The priest sighed.

I opened a pack of cigarettes, and they both took one. We drove on silently. . . .

An ox cart crossed the square [in Sukabumi], the driver sitting half asleep on the shafts until people began to shout at him and point upward at the helicopter, and when he understood he jumped down and ran to beat the ox with his hand but the beast shook its head and plodded on without changing its gait. The children had come from under the trees and were standing with their faces turned to the sky. A waiter, coming from the kitchen, walked out into the square with a bowl of rice in one hand and cutlery clutched in the other. People in the cafés, left their food to the flies and came outside into the sunlight, looking upward.

A driver shouted to a boy on the roof of his bus, asking him to hand down baggage; but the boy, gazing into the sky with his mouth open, bag in hand, did not hear but suddenly looked down and shouted, 'It's the President, the President.'

The helicopter was dropping down. The goats, alarmed by the loud noise of its engine, lifted their heads high and ran with uncertain steps, bleating. A boy threw stones to chase them to safety. People were coming from everywhere, out of side streets into the square, lining its sides, eager, excited. The helicopter steadied and hesitated a few feet from the ground, its rotor beating up dust; then it settled gently in the center of the square and tick-tocked to a stop. Hundreds of men, women, and children stood around the sides of the square, watching as if hypnotized.

The door in the side of the machine opened and a young soldier jumped out, wearing the red beret of the Palace Guard. He let down a set of folding steps and stood back, saluting. There was absolute silence, then a sigh like a wind stirring trees, and the President stood framed in the open doorway like a statue. It was theatrical and eerie. The whole town seemed to catch its breath, to gasp and become perfectly still for several seconds; then there was a scampering and whispering that reached into side streets and alley ways as more and more people began to stream into the market place. The whispering grew to a sea sound with shouting and the high-pitched cries of children excited and partly frightened.

The President stepped down and came toward the trees where the children had been playing: tall, straight-backed, impressive; aware of himself and of his powers; playing a part, creating an image of greatness. The young soldier hesitated, turned to follow, then turned back again to the helicopter and reached up a hand to help a woman down the steps. The crowd spoke her name, 'Hartini'. She walked after her husband smiling, bright-eyed and very beautiful. A few other people came after her: a young man who may have been a secretary or aide; two white men who had the surprised, uncertain, bemused look of visitors caught up in something unexpected; then two young palace guards without arms.

He came into the shade of the trees and stood waiting for Madam Hartini and her companions while men, pushing through the crowd, brought chairs for them. Within minutes, it seemed, there were perhaps ten thousand people packed in the square staring at their President and his lady, whispering, nudging each other, standing on tip-toe, straining to see better. (At the café we stood on chairs and saw clearly above the heads of the crowd; other people were on tops of buses or looked out from upstairs windows.)

He wore a dark khaki uniform, the ribbons of the revolution, the familiar black cap, and he carried a baton. He sat looking around calmly, spoke briefly to his visitors, smiled at his wife. The townspeople in the front row, closest to him, looked awed and uncertain and were worried by the pressure of the crowd pushing them in toward the President.

Leaning forward a little he began to sing a Javanese song, beating time with his hand and encouraging those closest to join in. The crowd took it up but was interrupted by shouting and confusion in one corner of the square where town police and other men were forcing a way through. When they came into the little clearing in front of the President and had formed up in front of him, he held out a hand to greet them: the Bupati, the mayor, the chief of police, the army commander, and the judge, still straightening clothes changed hastily, making it plain that the visit was unexpected.

These greetings made, they stood to one side. Women in the crowd began to sing again, with the men joining in where they could, and the President, the Bung, their brother, their uncle, beating time. He called two little girls from out of the crowd, who hesitated shyly but were pushed forward and in a while, with encouragement and cajolery, sang sweetly although petrified with fright and excitement; when they were done he called them closer and put his arms around him and spoke kindly before letting them go.

I looked at the people close by and saw in their faces such devotion and joy that I was amazed: men and women, faces alive with delight, some crying with happines, some praying. Whatever doubt I might have had about the power of this man, the gift he has of taking hold of the hearts of his people, the bonds of loyalty that bind them to each other, left me then. I could not believe it to be the evil fascination of a ranting, fanatical demagogue, but the genuine and rare power of a man who has the elements of both greatness and simplicity.

I, a Westerner, might feel afraid that this man can, and in my opinion does, make grave mistakes in the exercising of his greatness; but to his own people he is almost supernatural, a character out of the Wayang, who comes from heaven in a helicopter, sits before them like a sultan, and tells them what to do.

So now, in the market place, he called the five officials to ask if they had settled the troubles that had recently caused anti-Chinese riots in the town; had they repaired the damage and paid compensation to merchants whose stores had been broken into and looted? While they talked the crowd began calling for a speech and men attempted to assemble a public address system; the air was filled with the crackle and screech of amplified feedback and oscillation, which irritated him, for he waved the men and their apparatus away and called upon someone in the crowd to lead more singing.

Then he stood and began to walk back to the helicopter, the crowd parting to let him through, and as he went some drew back in awe and others leaned forward to touch, while a small boy walked behind staring up at him; Hartini followed, smiling.

The door of the helicopter closed on them. The engine spluttered and the motor began to turn, stirring thick billows of dust and grit. People moved back a little but continued to stare in silence until the machine, howling like a hurricane, rose off the ground. Then, even over the sound of the engine, I heard the crowd sigh as though released from an emotional imprisonment, and ten thousand faces stayed lifted up until the black dot disappeared, leaving an empty sky and silence.

Maslyn Williams, *Five Journeys from Djakarta*, New York: William Morrow, 1965, pp. 28–9, 37–9, 47–9, 263–8, 290–5, 338–41.

31
Eliot Elisofon Shoots Udjung Kulon, 1967

On assignment for Life *magazine, photographer Eliot Elisofon in 1967 spent three months in the Udjung Kulon wildlife sanctuary in the extreme south-west corner of Java, home of the elusive Javan rhinoceros (*Rhinoceros sondaicus*). Here*

Elisofon describes getting started—like many journeys to Java, his began with an encounter with the bureaucracy—and recounts one of his first treks into the park. He also tells us what the well provided for, modern-day photonaturalist brings along for a lengthy stay in the wilderness. (Accompanying Elisofon in Udjung Kulon was Lee M. Talbot of the Smithsonian Institution.)

WE flew directly to Djakarta. The only event was at Singapore, when we picked up a beautiful Swedish kerosene refrigerator, which was waiting on the field as luggage, courtesy of Peter Simms, the *Time-Life* correspondent there, who had answered my cabled query about these machines. To say that I surprised the customs men in Djakarta with a crated $4\frac{1}{2}$-cubic-foot refrigerator in my luggage is to put it mildly. Talbot and I cleared customs quickly though, and then were met by very friendly officials. In a few hours we were comfortably ensconced in the Hotel Indonesia in Djakarta.

Saturday, August 5, 1967, 10 p.m.

We are on board the motor ship *Samudera*, a 200-ton diesel vessel belonging to the Marine Research Department of the National Biological Institute of Indonesia. We left Djakarta about midnight today, or maybe it was still August 4—but this is trivial after five incredible days of Southeast Asian diplomacy in the capital. Three different groups have been wrestling for control of the country's science programs, including conservation, and Talbot, who knows them all, steered me from one contact to another, keeping each one happy.

Our being on the *Samudera* is not entirely by choice. A letter from Charles A. Lindbergh to Sultan Buwono, the third-ranking cabinet minister in the government, had stirred up great activity in Djakarta in anticipation of *Life*'s visit. At my first meeting with Dr Suhadi, head of the country's science programs, I was surprised to discover that his department had made an elaborate plan for my coverage of Udjong Kulon. To begin with, I was to use the *Samudera* as a base camp offshore. Next, six Indonesian scientists—the departments of biology, zoology, marine biology, forestry, and game were all represented—had been chosen to help me do my story. In addition, the government's official tourist agency, Nitour, had appointed a safari manager, Mr Sugiyono, to provide every possible comfort for the expedition, and Mr Sugiyono, I learned, had hired a chef and two assistants from the Hotel Indonesia in Djakarta.

The *Samudera* was to cost $250 US per day, running or not, and the safari $25 a day per man. For a two-month stay in Java the cost would be about $30,000.

After my shocked reaction Dr Suhadi brought out a $10,000 figure based on a shorter stay and a smaller group....

Talbot and I estimated that we'd need about eight men: a camp manager and a cook, who would stay in base headquarters once it was established, two more men, preferably with experience in the Game Department, who knew the animals and could track them, and about four men to carry the cameras and camping equipment on the many sorties we'd have to make to cover the various parts of the sanctuary. All this was agreed to in the presence of Sugiyono, the safari man, who became sadder and sadder as the prospects of his becoming a Nairobi type dimmed out altogether.

The upshot was that we were to get a boat for about $25 a day, as against $250, and could support our eight-man crew at about $1 a day each for food and about $10 a week each in salary, as against the $25 a day each it would have cost for the scientists' upkeep plus salaries....

I chartered the *Samudera* to bring us to Udjong Kulon as a small face-saving gesture. It had been standing by for us—or so, anyway, I was told—and I thought it would be polite to at least allow the Marine Research Department to earn $500 for the two days it's supposed to take the *Samudera* to get us to Udjong Kulon and itself back to port....

We left Bogor in good spirits, having ordered the *Samudera* to be ready on August 4. This gave us a day before sailing to buy food for the expedition. It must be remembered that once we left Djakarta, we left behind all adequate sources for provisions. There is an international food and appliance store in Djakarta where all purchases are made in US dollars. Diplomats and press people do not pay local tax. I stocked up on what I thought might be needed for at least two months in the jungle, and this filled ten large cartons.

The *Samudera* was supposed to sail at about 5 p.m. yesterday, but the captain had not obtained the proper clearances to leave Djakarta when we arrived at the port, and since yesterday was a Friday, a Moslem holy day, the port officials could not be found. We were told to wait until Monday, for all offices were closed over the weekend. But with us to cope with problems like this was Sri Budojo, an assistant of Sultan Buwono. Sri Budojo can only be described as a real operator; he could fix anything. It took him five hours, but he found the captain of the port and another functionary whose signature was necessary—and off we sailed.

To our surprise, we discovered Mr Sugiyono, the safari man, on board the *Samudera*, with a photographer from the Forestry Department as well. Both carried full camping gear. . . .

I didn't want amateurs in the area I was going to work in I told him, and after explaining—at length—I convinced him, then the Forestry man, to go back on the *Samudera* when she returned.

The *Samudera* stopped at Labuan in the morning, a small seaport four hours west of Djakarta by car, and here with the help of Mr Djuhari, the game warden in charge of Udjong Kulon, we found the *Harini*—the name means 'today' in Indonesian—and made a quick deal with the owner's agent, Ong Tjinbun, a Chinese Indonesian, to charter her and crew. The *Harini* is to follow us to Udjong Kulon.

At Labuan we also bought a large dugout canoe to use in ship-to-shore operations, and we hired and took on board the *Samudera* Amir Hasan, a young, educated clerk in the Forestry office at Labuan, who speaks a little English. Amir will be our camp manager.

Our next stop was at Tamandjaja [Tamanjaya], the nearest village of farmers and fishermen, on the north shore of the western end of Java, just before the neck of land which opens up into the Udjong Kulon sanctuary. We went ashore in the ship's boat and chose the men who'll be picked up by the *Harini* tomorrow. While on shore, I bought a dozen chickens, a small goat, and a pair of Muscovy ducks.

The *Harini* isn't a fishing boat; it makes its living running freight and passengers from Labuan to Tamandjaja and points in between—there are no roads west of Labuan, so all transport is by sea. I asked Ong Tjinbun, the agent, to procure fishing gear and put it aboard the *Harini*, since I expect the sea to be our main source of fresh food. Feeding the crew isn't going to be a problem. They eat rice three times a day, augmented by noodles, peanuts, beans, and any vegetables they can get, like cabbage, onions, or carrots, and finally either fresh fish or dry salted varieties. Fish about the size of an index finger can be smelled and seen drying on the beaches of every fishing town of Java and are the country's major protein.

In Djakarta I'd bought plenty of canned corn beef, Spam, and frankfurters, pork and beans and chili, tuna fish and sardines and tomato herring, fruit and juices of all kinds, and such canned vegetables as beets, string beans, and asparagus, which are excellent as salad. I also bought olive oil, vinegar, and garlic, as well as jars of mayonnaise. I didn't forget powdered milk and Quick Quaker Oats; this, cooked together, makes a good breakfast to start any hard working day. I have brown sugar and honey for sweetening. There are canned butter, peanut butter, jam, and cheese from Australia. I also bought packages of egg noodles and spaghetti, since I expect I'll get tired of rice. Funke's canned pumpernickel (from Holland) may prove invaluable, for the Indonesians are not bread bakers (the safari cooks in East Africa make great bread every day on the camp-fire); I also have several boxes of Bisquick and might try that if desperate. Of course there's tea and coffee, and there's Rose's Lime Juice and other flavors for the boiled water—flavors like bourbon and Scotch. I even have a case of wine, since I believe nothing is too good for the working class....

Talbot and I each have a small cabin—one bunk, a sink, and a porthole. Last night I discovered the sheet is too short for the bunk. It kept crawling up, which is annoying anywhere. Talbot was annoyed by this too. He was also upset because the *Samudera* had been fitted out as a marine research vessel by the World Wildlife Fund, and the laboratory on board has somehow disappeared....

The idea of a list of what I brought here still fascinates me. There is always one haversack, the kind you sling over your shoulder; mine dates from World War II—the North American campaign—and came from Abercrombie and Fitch. It is still solid, albeit dirty. I call it my war bag, and it has gone to

Africa, to New Guinea, and on every bush trip I've ever made. It contains:

1. A small kerosene Primus stove in its own tin box. (You can find kerosene everywhere in the world because it's what's used for wick lamps.)
2. My old Boy Scout mess kit in a canvas case. Folding frying pan, tiny pot, a plate, and a cup. With the mess kit, a magnificent steel gadget consisting of a can opener, knife, fork, and spoon that nest together. This has a swastika on it, which isn't surprising, since I took the gadget from a burned out German tank at El Guettar. There's a cheap copy of it made now.
3. A flat one-quart canteen.
4. A cot-sized mosquito net with nylon strings to hang it from. . . .
5. A sewing kit that includes a stitching awl and waxed thread; also, patches and glue for the air mattress.
6. A first-aid kit, including a snakebite outfit. The kit varies in size from expedition to expedition. For Java, I brought everything: Achromycin, aspirin, codeine, Miltown, Halazone (purifies any water), Mercurochrome and Merthiolate, thermometers, forceps, needles, surgical floss, bandages, swabs, tapes, sprain bandages, Band-Aids (most essential), gauze pads, ointments and salves, sunburn lotion, Alka-Seltzer and Bisodol (for results of my own cooking), Lomotil and paregoric (for diarrhea), antimalaria pills, potassium permanganate. . . .
7. A small bug bomb and its companion, a small bottle of insect repellent; a two-cell flashlight with two extra batteries; and a half-roll of toilet paper. A full roll is too large to include in the bag. I never, never go anywhere except East Hampton without No. 7.
8. A fairly large and heavy Sheffield steel sheath knife, which can double as a machete or a small ax. A tiny Carborundum stone for the knife and a tiny can of machine oil for the stone (and for guns).
9. A thin nylon cord that will support a man.
10. Fishhooks of different sizes and a monofilament line.
11. Sometimes the Walther .765, with extra rounds, and also, its cleaning rod and patches.
12. A pint flask for whisky.

13. A Boy Scout-type knife and an army marching compass, both on lanyards and worn on the person during marches; a waterproof matchcase; and an Abercrombie and Fitch altimeter.
14. Tiny five-power Japanese Nikon binoculars, which weigh no more than book-author-type horn-rimmed glasses. I also have ten-power binoculars with me, but I don't carry them in the war bag because they're too tricky to spot with—the game, particularly if it's a bird, is sometimes gone before you find it.
15. In cold climates, I manage to stuff in a knitted helmet, some wind goggles, and a pair of gloves. I also have army-flyer-type sunglasses. . . .
16. Sometimes a thick candle.
17. Sometimes a steel mirror for signaling.

For clothing on this trip, my basic outfit:

1. Three pairs of khaki trousers without cuffs (they catch in everything and you can't tuck them into socks).
2. Three long-sleeved khaki shirts, three T-shirts, and three pairs of boxer shorts. Six handkerchiefs (I'm using colorful Western-type ones here, because they're large enough to fit well under the neck of a shirt and keep ticks from crawling in).
3. Three pairs of reinforced athletic hose—normal length; one longer pair for tucking in trousers.
4. An old pair of bird-shooter boots, half-leg high for maximum support and snakebite prevention; a pair of Marine-issue reverse leather boots, ankle high (no desert boots, which wouldn't last two weeks on coral rocks, mud, etc.); a pair of khaki basketball-type sneakers (from L. L. Bean), which has proved great. The sneakers dry out quickly (far more quickly than leather footgear would) after wet-foot landing from the canoe to the beach. Happens all the time. Also, sometimes you just have to cross a shallow river or a swamp. Another thing to the sneakers' credit: good rubber bottoms, thick enough to stop most thorns. But not all of them—I've pulled plenty out of my feet. The lower stem of the "salah" palm here is so thickly studded with solid two-inch thorns you can hardly see it. If you step on one of these thorns, which are often hiding below dead leaves, you're really stepping

on something. When your feet stop missing those thorns, and get you fouled up in roots all the time, it's past the hour when you should be marching here; you need a rest, and as soon as you can get it.

5. A denim jacket and a sweater (I have yet to wear either of these here, where it's cold in the middle of the night, but not before). A plastic raincoat and a matching hat. Also a lightweight brimmed all-round hat (from Lock's in London) and an Israeli army hat. A pair of slippers and a Japanese cotton robe.

I also have two large towels, an air mattress, sleeping bag, air pillow, two sheets, a lightweight cotton blanket, and two pillowcases.

Just remembered I didn't list my toilet kit for the war bag. Also in it are a bar of soap in a plastic case, and a can of flea powder. (I'm allergic to flea bites.) And something I forgot to list with the contents of the medical kit; some antibacterial, antilouse soap. . . .

It might be of interest to list the lenses I brought with me to Udjong Kulon, for my four Nikon F (reflex 35mm) cameras.

21mm F4
28mm F3.5
35mm F2.8
Short 42mm–86mm zoom F3.5
55mm Micro-Nikkor for close-ups
58mm F1.4
105mm F2.5
200mm F4
300mm F4.5
500mm F5
1000mm F11

The lenses I usually find most useful are the 28mm, short zoom, and 200mm. These I have in a small camera bag with two of the cameras, a Weston exposure meter, and a bunch of film. The rest are in another bag, to be carried by one of the trackers.

I have several hundred rolls of film, mostly Kodachrome II and some high-speed Ektachrome; most of this is stored in the refrigerator until needed. But you can't use film that's just come out of the cold, because moisture condenses on its surface; you must wait after four hours to be safe.

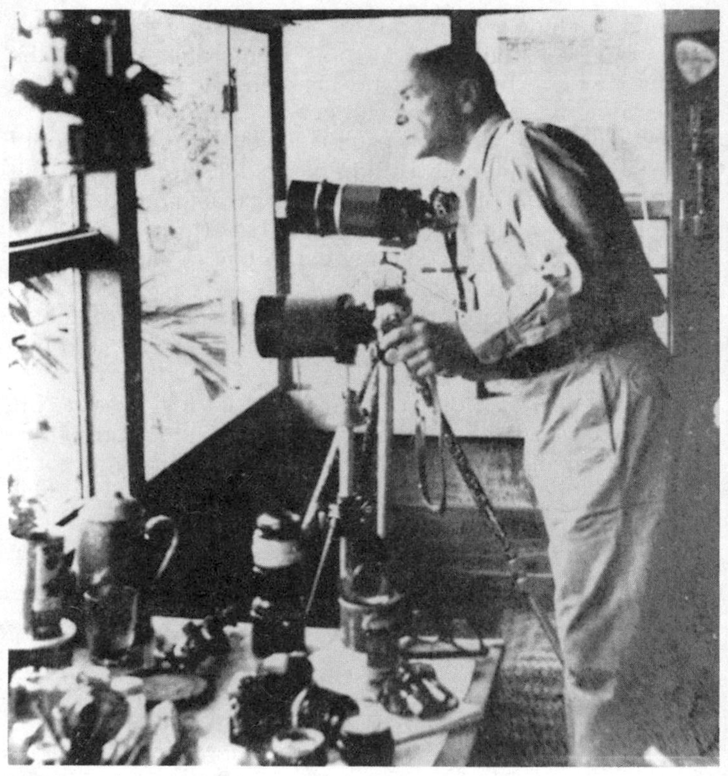

Photographer Elisofon and his cameras, Udjong Kulon, 1967. (Reprinted with the permission of Simon and Schuster from *Java Diary* by Eliot Elisofon. Copyright © 1969 by Eliot Elisofon. An Original Macmillan Text.)

More important, exposed color film must be kept cool until developed.

I have two tripods, both Quickset: a light one for portability and a heavy one for stationary setups, like the tower window. I also have two small strobe lights for interiors, balancing hard sunlight outdoors on close-ups, and the like. I brought two Weston meters and two Lunasix ones as well, and a Pentax spot meter. I also have a small underwater outfit: the Nikonos camera designed by Cousteau and a Sekonic underwater exposure meter. And I brought twenty-four cans of silica gel to dry the moisture out of my cameras and lenses. I don't have a developing outfit with me to make tests, but I'll be able to determine exposure meter accuracy by trial shooting with the roll film Polaroid camera I did bring with me. I have one more camera, a 35mm Widelux, a panorama camera with a view of 140 degrees. I also have a tele-extender, which doubles the focal length of any telephoto lens....

Sunday, August 13, 8 a.m.
Our survey team crossed the peninsula from opposite Handeuleum island and went westward and then southwest for about two-thirds of the way, then changed to northwest to reach the tower. This was all done on pure compass direction, since there were no landmarks or trails to follow. A major problem was how far we had gone. Checking my altimeter with the height—shown on our large-scale map—of various hills we crossed gave us some orientation. But difficulties were caused by the map, which turned out to be very inaccurate, especially in the placement of rivers. Often a river was indicated, but we found none; of those we did find almost all were dry, waiting for the rains to come to activate them once more. The dry river beds had cut shallow ravines in the forest floor; most still had some mud in places. The Tjigenter, the mouth of which was our starting point, kept crisscrossing over our due-west route and became a terrible nuisance.

On the very first day, I suffered a small embarrassment with the group. Most of the men had walked across the Tjigenter on a dead tree that had fallen lengthwise over the river's bed. I followed, proceeded a few wobbly feet, then suddenly thought, what if I fall off and break a leg? I had two Nikons to protect, and because I lost my nerve, I used them as an excuse for sitting down on the log when I was partially across. Without much shame I sat down and shimmied my bottom across. Ignominious, but practical.

We didn't get very far the first day, since we didn't start till midafternoon. It was good to let my muscles out a bit. Talbot's just as tough as the Indonesians, but I'm not.

Our walking order consisted of Harum as lead man, then Talbot, then Enang and me, followed by the rest of the men; Sinaga was usually at the end, keeping the column together. Enang had the larger Nikon Halliburton case, which resembles a one-suiter, strapped to a metal-frame pack carrier I had brought with me for that purpose.

The first man and sometimes Talbot did the chopping. Without the machete going all the time there was no way to get through this jungle. Udjong Kulon is a typical low-lying tropical rain forest, with dense foliage, all of it primary (but most of the really original flora was swept away by the Krakatau tidal wave). No one has cut it. This foliage consists of large trees, smaller ones, bushes, and many varieties of vine and rattan. The vines vary from finger thickness to that of a man's thigh; the rattans come in innumerable species, all of which are equipped with murderous thorns and barbs—not simple thorns, but bent ones, curved like fishhooks and placed most thickly not on the trunk of the plant, but on long whiplike ribs extending from the end of each leaf. The rattan actually climbs to the sun as the fronds catch successively higher branches and keep growing upward out of the deep forest shade. The barbed tips are very delicate and sway gently in the wind; unless severed by the lead men, they think nothing of catching on your face or neck and tearing out a piece. When one hooks you, you stop dead, then backtrack and try to disengage the barb. Doing this all day is no pleasure. Some of the other plants had thorns, but they were nothing compared with the rattans.

Once we left the flat land of the narrow coastal belt, we began going up and down. The ups and downs were never so abrupt that what we did could be called climbing, but they were steep enough—especially since the ground was made very slippery by millions of decaying leaves. The age-old habit of grabbing a tree branch or bush to hang on to was quickly cured with the first thorns. It seems that everything that grows here has sharp defenses. . . .

Tuesday night we made our first camp, about two kilometers, a little less than a mile, from where we'd started out. We were near a dry river bed, which we hoped would yield enough water to cook our dinner. It didn't, but in less than half an hour all the men who'd gone out scouting for water came back with armfuls of heavy vines. These vines are as thick as your arm; rooted in the ground, they climb a hundred or more feet up the trees to reach the sun. Water goes from the ground to the top, and the entire length of vine is loaded with it. The water has just a trace of sweet taste to it. By cutting short, pointed lengths of vine, the water can be made to drip out at an even rate. One gets about a half pint of water from about four feet or so of vine. We drank by holding the vines over our open mouths, the way Spaniards hold their wine jugs, but I didn't find this too easy. Some of the vine water—or juice—we used to boil our rice for dinner. The only water we carried on the survey was in two small canteens I had with me, and I usually filled one with tea. This improved the taste of river water that had been boiled, or if not boiled, doctored with Halazone water purifier tablets.

The men also cut many fronds of arenga palm, the single most useful plant in Udjong Kulon. Arenga palm is called *langkap* in Sundanese, the language and people of West Java, but game wardens have learned its Latin name, *Arenga*; its full name is *Arenga obtusifolium*. It has large leaves, like the coconut, and most important, no thorns. This makes it easy to handle. The men brought in dozens of branches and spread them out to sleep on. I thought they'd weave some of the leaves together to make a roof, but because it's the dry season, a roof wasn't necessary. Each man slept in a cotton blanket of some sort, in which he completely wrapped himself, covering even his head. I was shocked the next morning to see what looked like a row of mummies near the fire, which had been kept going during the night. . . .

Wednesday morning, about an hour after we started out, Talbot, who has eyes like an eagle—nothing escapes him—spotted a fresh rhino track in one of the still wet portions of the Tjigenter. Here was my first encounter with the 'badak', as Indonesians call the rhinoceros.

Eliot Elisofon, *Java Diary*, London: Macmillan Company, 1969, pp. 20–4, 26, 28, 119–22, 31, 33–4, 43, 45, 47–9.

32
Christopher Lucas Says Java is 'A Happening', 1970

'Indonesia isn't a country,' writes Christopher Lucas, reflecting the zeitgeist, 'it's a happening. If the hippies only knew, it's the ultimate total experience.'[1] *To him, Java seems like one huge improvisation, something like his trip. By the time Lucas makes the scene, Sukarno has been rendered laughable and the generals are firmly entrenched under the new leader Suharto. And even though Jakarta is still an eyesore full of 'murky slums, and other sores of poverty', a period of 'house cleaning' has begun. In the following passages, Lucas checks out the local colour in Jakarta, takes the train to Yogya, and then motors with the protean Chipto to Surabaya.*

I'VE been chewing my nails for six days in Djakarta, waiting for a general who promised a car. Within hours, they keep saying, I'll be sweeping through the scenic wonders of Java, en route for Bali. But my general has flitted to San Francisco, his aides have vaporized, and every other general has either gone to Buenos Aires, Milan, Tokyo, or simply isn't there. I spend my mornings drifting through bleak offices where the clerks smile sadly. In present-day Indonesia, it's worth knowing, Generals are the only people who get things done, and motor cars are scarcer than hens' teeth. 'So why not play golf?' they keep asking.

It's hot, humid, and my shirt is soaked. I've got stomach cramps, insomnia, and I'm getting nowhere. I'm bogged in promises, up to my eyeballs. And I wish I'd never come.

Regrettably, Djakarta is the unavoidable port of entry into Indonesia. As Paris isn't France, however, Djakarta isn't Indonesia, which is fortunate. For the capital is a fiasco. Its steaming, scruffy Kemajoran Airport is probably the most depressing on earth. There's nowhere to sit, the walls perspire, confusion is total, and the more timorous visitors get urges to board the next outgoing flight. And the city itself, at

[1]Christopher Lucas, *Indonesia is a Happening*, New York and Tokyo: Weatherhill, 1970, p. 16.

first glance, seems to offer all the more dubious exotica of Calcutta or Mao's Canton.

Besides malodorous canals where naked boys paddle, besides corrugated boulevards, dust-caked palms, and the stifling heat, Djakarta also has an exploding population, murky slums, and the other sores of poverty. From horizon to horizon, nestling between clumps of bamboo and blazing flamboyants, the flaking shacks and houses, still implausibly roofed with ruddy Dutch tiles, stretch to eternity. 'It's a village which has split at the seams,' says a Dutch old-timer. 'Four million people in search of a city planner.'

After a few days' exploration, however, Djakarta does hold a certain perverse fascination....

It's a disquieting sensation. My first morning, I wasn't sure whether the city had been hit by a hydrogen bomb or a building boom. Actually, it was neither. Across Djakarta, towering steel skeletons, their girders rusting, prod the sky; whole acres of cement and empty scaffolding stand ready for hotels and office blocks; the extravagant framework of a vast mosque looks particularly forlorn. These are the half-completed follies of Sukarno, the fallen demagogue, whose more practical heirs have blocked all his swollen-headed schemes and prefer to spend what's left on filling potholes and balancing the nightmare budget.

After two decades' neglect by President Sukarno, who squandered billions on colossal stadiums, conference halls, and Soviet-styled statuary, the once-trim squares carved out by the Dutch are rank with weeds, scrawny goats and heifers graze between the neglected flower beds, and barefoot peddlers trundle their handcarts down crumbling, midtown sidewalks. At strategic corners, Sukarno's monumental eyesores—the workers rending their chains, freedom fighters locked in everlasting battle—still flaunt their Marxist, and distinctly obsolete, iconography. Biggest of all, the National Monument is a gold-tipped obelisk which rises 330 feet from an apparent soup bowl and is irreverently known as 'Sukarno's last erection'.

'Three years ago Djakarta was a dead city,' I was told repeatedly. 'Now at least it wants to live again.' Maybe. Djakarta today proudly boasts its first department store, the fourteen-story Sarinah, complete with an improbable roof-top casino.

Thamrin Avenue, the capital's Main Street, is finally being repaved, and even the British, whose embassy was fired and ransacked by Sukarno's hoodlums, have seen fit to finance a handsome replacement. I'd hardly call the mood optimistic, but the ghosts have been exorcised, and the house cleaning has begun.

There's an impressive new motorway which goes nowhere in particular, half a dozen new office complexes for government penpushers, and a mushrooming slice of suburbia known as Kebajoran Baru, where the bungalows are spruce and the lawns get sprinkled. Even the venerable Djakarta Museum, which owns a musty but priceless collection of Chinese ceramics, has been given a dust-up. Which didn't inhibit the watchman from telling me, one hour before closing, that I'd, please, better go. 'The museum isn't full enough,' he said.

After the xenophobic years of abandon, Djakarta has also won back the confidence to recognize and restore the heirlooms of Holland. President Suharto and his generals suffer no complexes about the past. Once the Dutch governor-general's residence, the gracious Merdeka Palace, with its fluted columns, colonial facade, and chandeliers on the portico, has had a fresh coat of paint, as have the stolid, Dutch-built ministries (run by Generals), the Dutch-built banks (run by Generals), and the better Dutch-built homes (run by Generals' wives).

All in all, there's a prevalence of Generals, and it's a fact of life that they run the country. President Suharto, who dumped Sukarno into political limbo, is after all a military man. He trusts the tools of his trade, and in a country where corruption is endemic, he also trusts his own men above others. Whether his confidence is reciprocated—well, that's none of my business. Unlike his predecessor, Suharto does, however, like law and order, and he makes his intentions both clear and visible. He has 360,000 troops fully mobilized, keeps armored cars and 50-mm cannon parked at every university campus, and when the fancy strikes him, rumbles a squadron of Soviet-made heavy tanks through town. As a reminder.

By my tastes, there are rather too many troops around for comfort. Yet the military presence isn't suffocating, just ubiquitous, and most Djakarta citizens, who remember the grisly bloodletting of 1965, accept it all with engaging fatalism. Like my friend Karim. 'Suharto? Sukarno? What's the damn difference?' he asks. 'The average Indonesian still earns less than thirty-four dollars a year and dies before he's thirty-two.' And as our springless old Hillman shudders into gear, the laconic Karim, a university lecturer with damp sideburns and a Zapata mustache, throws us into frontal conflict with Djakarta's traffic.

Chewing an unlit cheroot, Karim plunges straight into a swarm of *betjak*, the gaudy, painted tricycles that serve as taxis. The *betjak* scatter. We miss an oncoming buffalo cart by inches, swerve around a stalled bus, graze a cyclist who's balancing six terra-cotta pots on his head, jump an open sewer, and finally grind to a halt on the sidewalk. Karim barely has time to point out a grim, gray-washed fortress. Its high walls fairly bristle with broken glass. 'That's the jail,' he gasps. 'It's packed with Communists. Three years ago, it was packed with anti-Communists.' And we're off again.

With reckless bravado, Karim dodges through an impossible maze of narrow streets. The pavements are crammed with stalls, and in the swirl of dust and exhaust fumes, the owners nonchalantly build their pyramids of mangoes, wrap a stick of root ginger or cinnamon bark, pour a jigger of rice wine. A paratrooper dozes against a tree, his pink beret drawn over his eyes, impervious to the prowling dogs, the sweet-sour smell of cooking spices, the admiring glances of three dusky schoolgirls. A swaddled baby lies in a doorway, but he doesn't cry.

It's unclean, seething, confused. Yet there's no hint of India's squalid poverty. These people laugh, flirt, and walk with pride. They may have no money, but their stomachs are miraculously full, and for six days I have not heard a voice raised in anger....

As dusk falls, both the temper and climate of Djakarta cool down. The dust settles, the smoke-spewing buses vanish, tumult turns into tranquillity. Aimlessly we drift toward the great, sprawling market place known as Senen, where sparks from a thousand fires flicker into the night. Like Marrakesh's Djema el Fnaa, it's a spectacle staged by De Mille.

Acrobats tumble, jugglers toss flaming torches, and a loin-clothed skeleton is eating light bulbs. Another mystic is frying eggs on his head. Clowns with paint-streaked faces joke obscenely, gray-bearded storytellers spellbind their circles of wide-eyed children, while medicine men wrapped in orange turbans offer potions for every ill from tired blood to broken hearts. 'Hold on to your watch,' says Karim. 'Our pickpockets have clever fingers.'

The eddying throng moves quietly and without haste. There is no jostling. The bare feet pick their way unerringly between the trays of vanilla and sandalwood powder. Cross-legged on open blankets, the tradesmen ignore their clients, squatting impassively between heaps of coffee beans, garlic, and tight bunches of peanuts, between grunting stacks of palm-cord baskets with pigs inside, between bolts of cloth, singing birds, plump pineapples, caged fireflies, and cobras. Over glowing braziers, old crones hunch like witches, frying bananas and sweet yams, and overall, there hangs an indefinable, aromatic cloud. It's the smell of cloves, which Indonesians blend with their tobacco.

The midnight air is mellow, and Karim's eyes are roving as he admires the slim, undulating contours of Java's women, voluptuous yet modest in their ankle-length *kain*. 'Our girls ripen early,' he smiles, 'as do our fruits.' And without explanation, Karim spins on his heels, walks a block or two, then turns down a narrow, dimly lit alley. We sneak past a police box with a drowsing sergeant, and Karim's friends wink. Behind iron-barred windows, there's the unmistakable outline of a bed and naked female shadows.

We're in the quarter called Little Paris, and evidently here's one problem which Suharto's puritans have *not* solved. From a doorway, a scrawny arm strikes out and clutches my shoulder. She's about thirteen, rumpled, and chewing betel nut. Karim thrusts her away. 'Let's beat it,' he says abruptly. 'People get knifed down here.' So we go to eat instead.

Djakarta's night life, as can be seen, is hardly exhilarating. Being a strait-laced Muslim city, there are neither clubs nor bars, and the social whirl means ginger pop at a roadside stall. Yet Karim promises a final ace. So after dinner, we hail two passing *betjak* and get pedaled to the People's Entertainment Park, which despite its sterile name is pretty extraordinary.

For a start, this gamy collection of side shows has been set down in dead-center, midtown Djakarta, right next to Sukarno's National Monument. Like opening a burlesque show opposite the Lincoln Memorial. And as we enter, a steel-helmeted riot cop frisks my pockets, then curtly prods me inside with his submachine gun.

The People's Park somehow combines all the more lunatic qualities of contemporary Indonesia. Down a one-time triumphal avenue, Beatle-mopped teenagers now race their go-karts. Like berserk kamikaze, they jump the sidewalks, screech between the palms, chase down a walking victim, or play suicidal 'chicken', while across the way, a twenty-five-piece gamelan rings its gongs for half-empty benches. On the left we pass an open-air go-go joint. The wild kids jerk with epileptic abandon, and the barker promises: 'Don't worry about partners! We'll look after you!' Next door there's a birth control clinic. And a slot machine for condoms.

In a bamboo shanty, we find the casino. It's run by Djakarta's military governor. The top stake is twenty-five cents. A painted sign reads 'Check All Knives and Pistols at the Desk'. And two unsmiling MPs slouch at the door with bayonets fixed. The casino is flanked by the Baptist Mission. We stroll past a row of foul-smelling booths, where they're hawking fly-covered hunks of mutton and other delicacies. Behind, a billboard states simply 'Entero-Vioform Is Best'.

There's also a shooting gallery with apparently live ammunition, an insistent minority of pimps, hustlers peddling pot, and inside one creaking pavilion, an absolutely ravishing team of dancing girls. Their mascaraed eyes flash; lustrous, raven hair flows down their slender shoulders. Perfumed and provocative, they roll their hips and shamelessly ogle the customers. Karim laughs. 'There's only one drawback,' he says. 'They're all men. Djakarta is full of sister boys.' I decide it's high time to go home.

In Indonesia, things are seldom what they first seem, and this can prove repeatedly unnerving. So back at the hotel, I drop in for a nightcap, and as at all times, the bar is packed. Germans, French, English, Americans—they're drinking with the tenacity of despair. Disgruntled, cranky, they're businessmen who came to do business and, in most cases, didn't. And their frustrations seem as endless as they are repetitive.

'Nothing changes in this dump, mate. Except these days the Army gets *all* the ruddy cumshaw!'

'Christ! I chased a shipment of quinine for three months and finally discovered it didn't even exist.'

'Two bloody cops hijacked my car, and it's now being driven by a major in Bandung.'

'Those f—g Generals! I've hosted fifteen dinners and still have no contract.'

'Money has no meaning in Indonesia. They barter cassava for coconuts. That's all they understand.'

'The troops broke into the hotel kitchen again tonight.'

'These blasted people don't even know what time means.'

It's the other side of the coin, and by their own values, the foreigners are, of course, right. For time and tensions don't exist in Indonesia. And payola is a way of life. Clocks, even calendars, are purely decorative. Like promises, they serve no useful purpose. So after squandering a week, I capitulate, as one frequently does in this country. I tear up my tidy, ten-page schedule, forget both my general and his car, and catch the overnight train to Jogjakarta.

Most Indonesian railways might have been resurrected from D. W. Griffith's *The Great Train Robbery.* The engines have cowcatchers and bulbous smoke-stacks, and the grimy, wooden carriages look like shantytown on wheels. But the prestigious Bima, which rattles down Java's backbone from the capital to Surabaya, is Indonesia's first and only 'express', which means it can, when lucky, hit a respectable fifty miles per hour. Custom-built in the People's Republic of East Germany, the Bima is diesel-fired, has flush toilets. It also has bunks for sleeping and an air-conditioning plant which can't decide between igloo-icy and steambath-torrid. But I can't complain. On my trip we had only three nonscheduled stops. Twice for errant buffalo and once for an old lady asleep on the tracks.

And as we slowly draw out of Djakarta, nudging goats and roosters to safety, small children run up waving and throwing flowers. Picking up speed, we roll past the nation's most blighted slums, tactfully shrouded in bougainvillaea. Minute by minute, the hot, dusty, urban disorder thins out, dissolves.

Then suddenly, extravagantly, on every side, the countryside of Java explodes in all its green, green, matchless opulence. The contrast is breath-catching, almost shocking, a stunning coup de theatre. The green rice shoots sweep from horizon to horizon, gently rippling in the breeze, with green-drenched hamlets marooned like islands between the paddies. Majestic coco palms, forests of bananas, green thickets of bamboo. A luxuriance of mangoes, papayas, and breadfruit, cotton, tea, rubber, oil and sago palms, and everywhere the clear, sparkling streams of life-kindling water.

It's a scene so immodestly prodigal that it makes Tahiti look barren. It's the lush fertility of the jungle that has been tamed and put to use. Caressed by sunshine and monsoon rains, the soil is so rich that maize yields in seven weeks and a banana tree grows in ten. It's no wonder Indonesians don't know hunger. Yet mile after mile, the flooded paddies just mirror the clouds, uncared for by toiling peasants. Once in a while we pass naked boys with coolie hats crouched motionless across a buffalo's haunches, but otherwise the land is deserted. In Indonesia, men do the planting and reaping, but Nature is the best husbandman. She bountifully provides two, even three, rice crops a year. And I thought of India, where every grain must be coaxed with tragic concern. . . .

In Indonesia, you don't travel, you improvise movements. From Jogja [Yogyakarta], the road swings northeast to Surabaya, but the buses are booked solid for three weeks. So I hire a cab instead. My driver is a college lecturer who teaches chemistry and charges me double. Like all Indonesians, he's moonlighting. He also seems to confuse highway travel with a bullfight on wheels.

Hand on horn, Chipto hurtles through the world's worst traffic with an assurance that is terrifying. We squeeze crazily between groaning trucks and smoking buses, zigzag through swarms of *betjak*, horse buggies, and oxcarts, haphazardly scattering geese and cyclists, chickens and strolling peasants. On screeching tires, we happen to be charging through the most thickly populated area on earth, but Chipto is not impressed. And promptly ignores the only traffic light east of Djakarta.

Chipto's foot is glued to the pedal. It's neither the time nor the place for sightseeing. Still, I do have fleeting, white-knuckled memories of the twin volcanoes known as Merapi and Merabu, teak forests, big red banners which scream 'Down with Moral Decadence!', rustling acres of sugar cane and tobacco, and the ubiquitous jigsaw of rice paddies. Then in mid-afternoon, we hit the monsoon.

Within minutes, the dikes erupt. The rain squall hits our roof like machine-gun fire. A torrential river pours across the road. Visibility is zero, but Chipto keeps his foot on the gas. Skidding and sliding, plowing through clouds of spray, we dodge eerie phantom shapes, knock something or somebody into a ditch, and after ten hours and just 170 miles, reach Surabaya. Will I need him tomorrow? asks Chipto. I smile wanly and call for a beer. A big one, please.

Three hours later, I'm with Tjiptono. He's Chinese, twenty-six, married, and vivaciously charming. We're eating in the open air. Wood benches, dirt floor, and cats howling at every table leg. It's hot, hot, hot in Surabaya. Maybe 105 or more. We're drenched with sweat, and I've just mistakenly swallowed a chili that makes Tabasco taste like tomato juice. We're nibbling through a selection of *sate*, tiny cubes of chicken and lamb that get skewered, barbecued, and dunked in a peppery peanut sauce. There's also a dish of sheep's eyes that keeps looking my way. Still, the *sate* is delicious.

'Me doctor. Earn three dollars one month. Most people no pay,' says Tjiptono. 'But me also sell 217 article last eighteen month, and that make me plenty rich!' He laughs. 'Now get top rate. One dollar for story.' Unlike Chipto, friend Tjiptono has *seven* jobs. He's a practicing physician, aide to the governor, tourist guide, magazine writer, shipping clerk, AP stringer, and occasional chauffeur. He's also staggeringly enthusiastic.

Would I like to see the Heroes' Monument, which looks like Asia's biggest ballpoint pen? Would I care to visit the zoo where the crocodiles come from Surabaya's backwaters and the giant *komodo* dragon eats four dogs for breakfast? Would I like to see the Soviet-built cruiser, which hasn't put to sea for three years? Would I care to view the bull races on nearby Madura, where the yoked steers cover a hundred meters in nine seconds? Wouldn't I like to visit Surabaya's six department stores, where, despite all Indonesia's money crises, you can buy everything from rubber falsies to Levis, plastic hibiscus, 'Miss Dior', and postcards of Betty Grable?

Eventually, as happens in Indonesia, I was shown everything except what I wanted to see, which was the bulls. But tonight I'm stubborn. I want to see *kuda kepang*, East Java's notorious 'horse trance'. I'd watched a shoddy imitation in Jogja, which cost me twenty dollars and was a phony. I was both disenchanted and skeptical, but Tjiptono shrugs his shoulders. 'Jogja people topside no savvy,' he says. 'Now we go People's Park. You look-see.'

Christopher Lucas, *Indonesia is a Happening*, New York and Tokyo: Weatherhill, 1970, pp. 19–20, 23–30, 42–4.

33
V. S. Naipaul among Java's Believers, 1981

Trinidad-born and Oxford-bred V. S. Naipaul, the prolific author of fiction and non-fiction alike, embarked sometime around 1980 on an exploration of four Muslim societies in the simultaneous throes of material modernization and religious revival. Java was his last stop. Here he visited the island's famous pesantren, *or Muslim religious academies, and probed the spirit of the times with poets and intellectuals, and his young assistant and guide, Prasojo. In Jakarta, Naipaul resided at the Hotel Borobudur, then the city's standard for five-star luxury. (The old Hotel des Indes had been torn down some years before.) Both in the hotel and out and around the island, Naipaul finds much about Java disquieting.*

THE Borobudur Intercontinental in Jakarta changed its character at Christmas. The men from the multinational companies, and the foreign economists and advisers, left. Many of them were solitary, middle-aged men. Some went home; some went to the cooler hills or to the islands. The Borobudur offered cut-price holiday deals for local people; and the local people came, with their families; it was a recognized way, among the well-to-do, of spending the holidays.

Children ran up and down the carpeted corridors and played with the elevators. Nannies or ayahs, some of them barefooted, dandled babies. One Chinese family, doing the right thing for the holidays but not enjoying it, spent a whole morning sitting silently on the upholstered benches outside the elevators on the fifteenth floor. The head of the family, an old man with a ravaged face, wore a singlet without a shirt. From the fifteenth floor the black-haired heads in the pool with the rippling Borobudur design seemed unnaturally large, and (also because of their number) suggested tadpoles. One morning I counted sixty-three heads in the pool.

Simple pleasures; but they were feeding resentment. Resentment of Chinese; of foreigners; of people with skills Indonesians didn't have. Resentment, perhaps, of the skills themselves, and the new order they were bringing in, which no one yet fully accepted: new men, new status, new power, new money. Wrong men had money. Wrong men gave themselves feudal airs. Wrong people romped about the Borobudur and showed the other side of the new society.

'Cheap for you,' the girl said at the hotel shop, when I bought a bottle of port for the holiday. 'But not for us.' And her big smile—yet not her old smile, not the smile I knew—was chilling.

The feeling of wrongness was there. All that had been done during the fifteen years of peace could be ignored. The richer the country became, the better it was made to run, the easier it was for its creative side to be taken for granted, the easier it was for the new inequalities to show. And people could long for 1945, when everybody was equally poor and everybody had the same idea of what was right and wrong. In the town, as in the villages, every improvement made matters worse, made men more uncertain....

I went [to Surabaya] with Prasojo, a nineteen-year-old college student, and I could not have had a better companion. Prasojo had been to Arizona for a year on a scholarship given by the American Field Service. He spoke English well, with an American accent. He had greatly enjoyed his time in Arizona, had learned much, and remained so grateful to the American Field Service that he intended to give them part of the fee he was going to get from me.

I also felt that Prasojo wanted to give back to me, a stranger, some of the kindness he had received in the United States. For our trip he wore jeans with the AFS label stitched on the hip pocket. He was just above medium height and of Chinese appearance. That appearance was the subject of a family joke. Prasojo's father, a bulky man, undeniably Indonesian, would say, 'But, eh—how did I get this Chinese son?'

We took the Garuda air shuttle to Surabaya, on the northern coast of East Java. Mud tainted the coastline. The rivers were muddy wriggles in the green, overworked, overpopulated land. The land around Surabaya was a land of rice, the rice fields in long thin strips, easier that way to irrigate, but suggesting from the air an immense petty diligence.

The houses—as we saw later, driving inland from Surabaya—matched the rice strips. They were very narrow and went back a long way. The houses stood a little distance from the road, and the front yards were scraped clean, but shady. Banana trees grew out of the bare earth, and coconut trees, mango trees, sugar cane, and frangipani. The rice fields began directly at the back of the houses. During that drive we seemed to be going through one long village: Java here an unending smallness, hard to associate with famous old kingdoms and empires, a land that seemed only to be a land of people of petty diligence, the *wong chilik*, the little people, cursed by their own fertility, four million in Java at the beginning of the last century, eighty million today.

It was Prasojo who gave me that word, *wong chilik*, telling me at the same time that the word (though beautifully appropriate in sound) was both insulting and old-fashioned. It still mattered to some people, though, who were not of the peasantry, to have their distinction acknowledged. Such people called themselves 'nobles', *raden*, and used the letter R before their names. They also built houses with a special hat-shaped roof, a distinction I would have missed if Prasojo had not pointed it out to me, so squashed and repetitive and cozy it had all seemed: the red tile roofs, the walls of woven bamboo for the poor, concrete for the not-so-poor, the yards full of shade and fruit and flowers.

Windows were an innovation, Prasojo said. In the traditional Javanese house there were none; and, with walls of woven bamboo that shut out glare and heat but permitted ventilation, windows were not necessary. In the traditional house, light came through gaps in the roof. But concrete walls required windows; and I could see that glass louvres were fashionable among the not-so-poor.

Each little yard had its gateposts but no gate. The posts were of a curious design, with slabbed or stepped pyramids or diamond shapes at the top, the pyramids or diamonds sometimes bisected: concrete, but concrete clearly imitating brick. These posts, which at first suggested a single ownership of land and people, perhaps by some vast plantation, were in fact the remnant of the architectural style of the last Hindu kingdom of Java, the kingdom of Majapahit, which disintegrated at the end of the fifteenth century.

This was how the pre-Islamic past survived: as tradition, as mystery. *Indrapura*, 'Indra's City', was painted on the bus in front of us; and *Indra Vijaya*, 'The Victory of Indra', was on many shops. But this Indra was no longer the Aryan god of the Hindu pantheon. To Prasojo, as well as to the driver of our car, this Indra was only a figure from the Javanese puppet drama. Prasojo began telling me a local Muslim legend of the five Pandava brothers, who represented the five principles of Islam. And I don't believe Prasojo had an idea of the true wonder of the legend: the story he was telling me came from the ancient Hindu epic of the *Mahabharata*, which had lived in Java for fourteen hundred years, had taken Javanese roots, and had then been adapted to Islam. Prasojo, a Javanese and a Muslim, lived with beautiful mysteries. Scholarship, applied to his past, would have undermined what had become his faith, his staff.

And so we came in the late afternoon to the town of Jombang. It was where the famous old *pesantren* was. But Jombang, once we turned off the highway, seemed to be full of schools. There were scattered groups of chattering Muslim schoolgirls on the road at the end of the school day: little nunlike figures, with covered heads, blouses, sarongs. Where was our *pesantren*, and in what way was it different from these other academies? We raced back and forth, the driver behaving as though he was still on the highway; we penetrated murky rural alleys. And then we found out that we had passed it many times: it was so ordinary-looking, even with a signboard, and not at all the sylvan retreat, the mixture of village and school, that I (and Prasojo as well) had been expecting.

There was a fence. And behind the fence, rough two-storey concrete buildings were set about a sandy yard, which had a few trees. In the centre of the yard there was an open pillared mosque with a tiled floor just above the ground. Boys in shirts and sarongs were sitting or lounging at the edge of the floor and on the step, following an Arabic text while a sharp-voiced teacher, unseen, steadily recited.

We went past the newspaper board—in the open, with a wooden coping, and with the newspapers behind glass—to the office at the side of the mosque. There was nobody in the office. Variously coloured shirts and sarongs hung on the verandah rails of the two-storey buildings. There were boys everywhere, barebacked, in sarongs, with warm brown skins and the lean, flat, beautiful Indonesian physique, pectoral and abdominal muscles delicately defined.

They stared back. And then, gradually, they began to gather around Prasojo and me. When we walked, they followed. They became a crowd as we walked about the narrow dirt lanes and the muddy gutters between the houses at the back of the compound: hanging clothes or sarong-lengths everywhere, glimpses of choked little rooms (eight boys to a room, somebody told Prasojo). There was mud and rubbish outside the rough kitchen shed and the school shop; and over an open fire in the muddy yard one little saronged boy was scraping at a gluey mess of rice in a burnt saucepan. He looked up in terror, at us, at the crowd with us. Perhaps, I thought, all medieval centres of learning had been like this. . . .

It was noon, humid even in the teak forests. Prasojo fell asleep from time to time. He had the driver play pop music on the cassette player in the car; he slept to that. The land flattened, opened out to a wide plain with a line of blue hills on one side and high peaked mountains far on the other. It had rained; everything glittered. The green of the paddy fields was glorious. And against this green every touch of bright colour—in the dresses and sarongs of the people working in the paddy fields—was doubly glorious, reflected, with the sky, in the water. The rice grew in straight lines; different fields were in different stages of growth.

As the light changed, as the afternoon heat faded, Prasojo stirred and became alert again. Sleep had more than refreshed him: he talked poetically about the country through which we were driving. He had been educated near here. He spoke of the beauty of getting up in the morning while it was still dark and walking with palm-frond torches to the road to wait for the bus. He spoke of the 'dating' habits of the afternoon. Dating time was between four and six; there was nowhere to go after seven. The girl sat on the back of the motorbike with her legs to one side and held the boy around the waist: that was the recognized dating pleasure.

'This is the best part of the day,' Prasojo said.

The sun was red. The light was red; it came red through the trees, fell red on the road. A faint mist rose off the rice fields; the blue hills went pale; and sun and sky were reflected in the water of the rice fields.

'For us it isn't easy to be abroad,' Prasojo said. 'We get homesick.'

They got homesick for everything. For everything we had experienced that day, the freshness of the morning, the heat of noon, the relaxation and colours of the late afternoon. For everything we had seen on the road and in the fields: the cycle of the rice crop, the changing tasks, the men carrying loads in baskets on either end of a bamboo pole, the bicycle rickshaws, the horse carriages (different regions had different styles of carriage). To an Indonesian everything about his country was known; no detail of house or dress or light went unconsidered. Every season had its pattern; every day had its pattern. When Prasojo went to Arizona his first thought, waking up the first day, was, 'I am not in Indonesia.'

All this was drawn out of him by the fading light, the best time of the Javanese day. The road was full of people yielding to the pleasures of that time of day, relaxing, chatting. The horse carriages were busy. Boys and girls rode together on bicycles—Prasojo pointed them out to me. . . .

A week later I went back to Yogyakarta, to go with Umar Kayan to one of the villages below the volcano of Mount Merapi. Merapi had a long, easy slope; a wisp of vapour always hung about its cone, and sometimes the cone was lost in cloud. Lava made the earth rich. The wet soil that the men in the paddy fields ploughed deep with bullocks or lifted with mattocks was black and volcanic. The mud-walled rice fields came right up to the villages, so that from the air the villages—red tile roofs among green trees, shade in the tropical openness—had sharp, angular boundaries.

To enter one of those villages was to find more than shade. It was to enter an enchanted, complete world where everything—food, houses, tools, rituals, reverences—had evolved over the centuries and had reached a kind of perfection. Everything locked together, as the rice fields just outside, some no more than half an acre, fitted together.

Every house, with concrete walls or walls of woven bamboo strips, stood in shade; and every tree had a use, including the kapok, new to me. There were many kinds of bamboo, some thick and dark, almost black, some slender and yellow with streaks of green that might have been dripped by an overcharged brush. These bamboos made beds, furniture, walls, ceilings, mats. But rice ruled. It was the food and the cause of labour; it marked the seasons. In the traditional house there was a small room at the back of the pillared main room; this small room, in the old days, was the shrine-room of the goddess Sri, Devi Sri, the rice goddess.

Umar took me to Linus's village, and Linus went with us. Linus was a young Yogya poet whose only income so far was from his poetry readings. Linus was a Catholic; his full name (he had an Indonesian name as well, but he didn't use it) was Linus Agustinus. Linus's mother was a Catholic; it was to marry her that Linus's father had converted. They were a farming family. Linus's father was the village headman. Since the military take-over this had become an elected post, and the headman had to see that the government's projects were carried out—getting the farmers to plant the new rice, for instance. The headman, while he held his post, had the use of twelve acres of land; in Central Java that was a lot.

The village was off the road to Pabelan, and Umar knew the area well. During the revolution—the war against the Dutch—the Dutch had invaded Yogyakarta, and the revolutionary army had moved out into these villages. Umar was in the students' army at the time; they were billeted on the villagers.

I asked, 'Were you well organized?' Umar laughed at my question. 'What do you think?' It had been a time of chaos;

and it was hard, as it is in most places in a time of peace, to think of war in such a soft setting: such small and fragile villages, such vulnerable fields, requiring such care.

Linus's family house was of concrete, on low pillars, with a concrete floor. But it was of the Javanese pattern. There was a Catholic icon above the inner door. And on a wall was a leather figure from the Javanese puppet theatre: the figure of the black Krishna, not the playful god of Hindu legend, but the Krishna of Java, the wise, far-seeing man, and therefore a suitable figure for a poet's house. In the bookcase were Linus's books from school and the university, and also *The Collected Poems of T. S. Eliot*, a gift from the BBC: Linus had won second prize in a poetry competition sponsored by the BBC Indonesian Service.

Glasses of tea with tin covers on top were brought, and a plate piled high with steaming corn on the cob, and then a plate equally laden with a kind of roll. It was the Indonesian ritual of welcome, the display of abundance. It called for a matching courtesy in the guests. No one wished to be the first to eat or drink; and it often happened that the tea, say, was drunk right at the end, when it was cold.

One dish was brought out by one of Linus's younger sisters, a pretty girl of ten or twelve in a frock. Then another girl, much older, came out to look at the visitors. Her face was twisted, her teeth jutted; her dress hung oddly on her. Her movements were uncoordinated, and her slippered feet dragged heavily on the smooth concrete floor. She sat on a chair in the other corner and looked at us, not saying anything; and then, after a while, she lifted herself up and went out with her dragging step.

Some minutes later Umar said, delicately, 'You may have seen that sister of Linus's. She is not well.' She had fallen ill when she was young. They had taken her to a doctor, and the doctor's assistant had given a wrong injection, which had damaged her nervous system. So the house of the poet, the house of the village headman, was also a house of tragedy.

Linus's mother arrived: the woman for whom the father had converted, and because of whom the family was Catholic. Umar got up, with a definite stoop, and did a shuffle sideways, a big man trying to make himself smaller than the small woman. And a stream of musical speech poured out of both of them before we all sat down.

She was small and slight in the Indonesian way, and she might have passed unnoticed in the street. But now, detached from the Indonesian crowd, in her own house, and our hostess, her beauty shone; and it was possible to see the care with which she had dressed—blouse, sash, sarong (her daughters wore frocks). It was possible to see beyond the ready Indonesian smile (disquieting after a time) to her exquisite manners, and to see in this farmer's wife the representative of a high civilization. Her face was serene and open; she held her head up, with a slight backward tilt; her bones were fine, her eyes bright, though depressed in their sockets, and her lips were perfectly shaped over her perfect teeth. Her speech—without constraint or embarrassment—always appeared to be about to turn to laughter.

She and Umar talked for some time in this way, and it seemed they had much to say. But it was all part of the ritual of welcome, Umar told me later. They had used the polite Javanese language, which was different from the everyday language; and they had said little. Linus's mother had said that she had had to go to the school of one of her children to get the child's report; that was why she hadn't been able to welcome us when we arrived. She was ashamed to welcome people as distinguished as ourselves in a place that was hardly a house, was a mere hut. And Umar had been equally apologetic about our intrusion, which was perhaps upsetting the harmony of her household. That was how it had gone on, apology answered by apology.

One concrete thing had come out, though. Umar mentioned it afterwards, when we had left the house and were walking through the village. Linus's mother was worried about Linus. He didn't come to the village often; he stayed in his little house in Yogya; he wasn't married; and he didn't have a job. And she had a point, Umar said: Linus was twenty-eight.

I said to Linus, 'But isn't she secretly proud that you are a poet?'

Linus said in English, 'She wouldn't have even a sense of what being a poet is.'

Umar said, 'There is only one way Linus has of making her understand. And that is to say or suggest that he is being a poet in the classical tradition. But that would be nonsense. She would reject it as an impossibility.'

For someone like Linus's mother, living within an achieved civilization, poetry was something that had already been written, provided, a kind of scripture; it couldn't be added to.

But something was about to come up for Linus. A Yogya paper had asked him to do a cultural page, for twenty-five dollars a month. It was a short bus ride to Yogya, but the life that Linus was trying to make for himself there—poetry readings, newspapers—seemed a world away from the tight, rice-created village. . . .

It was only half an hour to Yogya. But not all could make the journey from village to town as Linus had done. Linus was privileged. He was a poet; he had a sense of who he was; he could be a man apart. Not many villagers were like that. They had been made by the villages. They needed the security of the extended family, the security of the village commune, however feudally run, however heavy the obligations of the night watch or the communal labour in the rice fields. For such men the villages were indeed enchanted places, hard to break out of. And if a man was forced to leave—because there simply wasn't the land now to support him—it was for the extended family—and something like the village again—that he looked, in the factory or the office, even in Jakarta.

V. S. Naipaul, *Among the Believers: An Islamic Journey*, New York: Alfred A. Knopf, 1981, pp. 356–7, 318–20, 327–8, 343–6, 350.

34
John and Iris Do Yogya, 1984

Recent decades have brought a swarm of new travellers to Java—young, restless, backpack toting Westerners who spurn luxury hotels and tourist haunts and seek instead the 'real' Java. Authenticity for them often involves a succession of nights in the island's ubiquitous losmen, *or cheap hotels, and a preference for buses and other forms of proletarian transport. Slim budgets tend to make such choices a necessity, in any case. As John Krich and his girlfriend Iris discover, 'tourist Java' is hard to escape, and the real Java is, alas, often all too real. Here is his tale of their stopover in Yogyakarta, titled 'Shadow Play'.*

IRIS and I were stalled in the bell-tinkling traffic of Jogjakarta. Another rickshaw pulled alongside and a zoot suiter in the carriage called out, 'Australian?'

'American.'

'America. Very good. Hello to you!'

'Hello.'

'Married?' He got right to the point.

'No.'

Our inquisitor disappeared, his lane of bicycle stalled. We could hear him berating his driver until he was once again pedaling parallel to ours.

'Friends?' the young dandy resumed.

'Yes.'

'You live in same house?'

'Yes.' There seemed little harm in the truth. It was making him so happy.

'In Indonesia, never!'

He dropped back, then surged once more into view, a persistent Kinsey researcher.

'You sleep in same room?'

'Yes.'

'In Indonesia, never!' He was incredulous. 'In the same bed?'

'Where else?' He couldn't translate that phrase, so before he fell behind for good, I added another 'Yes.'

'In Indonesia, never!'

In Indonesia, what then? So far, we'd found a country in stasis, a people in a political state of grace—where the military took full advantage of the masses' tropics-induced tolerance, where the snoozing, sensuous body politic had rolled over in its hammock one too many times, crushing alternatives. Where the ghosts of six hundred thousand slaughtered Communists lurked. Where jeeps and lorries, paregoric-colored convoys, made up the occupational force called the government. Where banks and barracks were indistinguishable. Where patronage flowed from neo-Fascist pillbox ministries, complete with Dixie colonnades, and even the cinemas were impregnable bunkers. Where whole pages of *Time* magazine were blacked out. Where children scorched bulging plantains over open fires, where fathers scraped about in sandals and bathrobes from sleep room to eat room to card room across cockamamie family courtyards large enough to be called hamlets. Where mothers fanned their ample Micronesian forms, their faces so round as to not possess shadow, adjusting their one-piece batik gowns on the way to fetch rainwater for lemonade and feed the parakeets. Where life looked idyllic, until all the evidence was in, and we realized these people had nothing to do.

Hong Kong's precise social clockwork had jammed with equatorial rust, strewing human coils and springs everywhere. While the most fawning Chinese might have called me 'mister', here I'd become boss. But I felt like a stumble bum. Crossing the hemispheres had turned low winter into high summer. Iris and I were now paying for our show of geographic nonchalance. If we could, we planned to stagger across the sauna bath of Java and jump in the ocean's cold shower at Bali. Yet we were so unprepared to feel this unprepared! O, Java! O, dread coffee bean sprung to indecipherable life! Was Krakatoa brewing again? Was that the source of this everlasting steam?

Iris and I had to adjust our thermostats, and our expectations. Indonesia was the first lesson in 'Seeing the World Through Asian Eyes'. Jakarta had proved a city hiding from itself, a collection of peasant towns divided into monotonous blocks. The impressive boulevards carved cordons sanitaires in neighborhoods of squatters. Close to the national nerve center of Merdeka Square—merdeka meaning freedom, freedom from the Dutch—the streets were unpaved and the shop windows empty. Restaurants ladled out 'Padang-style' curries to disguise the condition of what little meat was on hand. Though this was the capital of the world's fifth most-populous land, one third-rate tornado, one efficient collection agency could have boarded up the whole shebang.

Our first Asian train ride was not just measured in miles skipped, but in scenery skirted. Eight hours on a slatted, second-class bench turned inevitably into sixteen, but where were the volcanoes, the rain forests, the monkey habitats? Plains of paddy flat as Kansas hypnotized us, and except for the locust waves of oval-eyed boys who raced through the aisles at each stop—their cries too grand for their banana leaf catering—we saw nothing to justify our decision to take the journey in the day's heat. It was only at night that there was anything much to see and that was the one creature who made sure we could see it: the firefly. And not just a few, mind you, like those stray kamikazes we used to chase and lunge for in the cool twilights of our frigid North American youth. Our train passed through a miles long firefly tunnel, a throbbing cloud of loop-the-loop current. Pattern within strafing pattern, glow subsiding to glimmer replaced by glow again, these 'lightning bugs', as Iris called them, did seem to crackle, conduct, and drain the darkness of its power. Interrupted only by the black-out of banana groves, we sifted for hours through the rice fields' astonishing electromagnetic harvest. Outside, out of reach, we found our first show of Asian abundance. Nature's generator was doing just fine, even if the lights in our slumbercoach never came on.

Jogjakarta, the 'cradle of Javanese culture', had turned out to be one glorified main street, the Malioboro. It was combed in klaxon-ringing swarms by glorified bicycles, the becaks that hauled an upholstered, fringed passenger seat which doubled as the drivers' homes at night. How could someplace so ancient look so temporary? The most solid edifice, and the oldest was the prince's candi, or palace, a pavilion of spindly pylons. The local university was laundry-bedecked, its enrollment largely of chickens and roosters. Downtown 'Yogya', as we learned to pronounce it, consisted of two covered uneven boardwalks, that had the feel of a pacifist Dodge City. No poker parlors or saloons here, just tinder boxes containing batiked surfer wear, hippie hangouts serving fruitwhips beneath creaking ceiling fans, one drugstore for every sick citizen. There were certainly enough horse-sized antiflu capsules being displayed, but 'enough' was a word that Iris and I would soon stop having cause to use.

As for the loesmans, or 'guest houses', where the pedal cabbies took us, they were no more than designated corners of those sprawling atriums in which the Indonesians sat out lifetimes with a minimum of disturbance. The inns' plaster dividers made a gallery for pin-ups of American rock groups who were vapid enough for export: the Spiral Staircase, Suzi Quatro. Masons' mishaps let in malarial moquitoes like the *Titanic* let in water. A nail driven into the wall served as a closet. Sheets? Sometimes washed, always stained. Telephones? The whole province was lucky to have one. Brotherhood was the only item in great supply, which wasn't so bad, except that the main time it got called into action was over the toilet, the squatter, the flushless hole—where guests learned to love the look of the next guy's worm-riddled feces as they loved their own. And when hotels down the road beckoned us with flyers promising 'MUSIC IN EVERY ROOM!' we were amazed when a single radio rasped in the common courtyard.

It was enough to turn anyone into a tourist. Even the all-natural Iris, who grudgingly conceded that where the present looked so unmomental—and the future positively eager for corruption—the best way to make contact with something we could label 'Indonesian' was through the monuments of the past. She was the one who usurped my guidebook and proposed an outing to the twelfth-century temple of Prambanan, some twenty minutes ride from town. Before we'd reached the bus station, another of the Malioboro's charming sycophants was at our heels. Like other practitioners of his amorphous trade, he made a dashing figure. The padded shoulders and baggy cuffs of the white linen outfit they all favored did a particularly good job of highlighting his long legs and sprightliness. Wavy black hair streamed down the back of his collar. A flowered tropical shirt went underneath, another Carnaby Street hand-me-down that seemed to find its best use here. The popular fashions were at least straining to be up-to-date, even if the popular history wasn't.

'First time in Yogya?' he called. 'You would perhaps like to come with me?'

'Where?'

'Where? Why, all about. On seeing of sights. Prambanan and Borobodur temples.'

'No, thanks,' Iris fibbed. 'We're leaving today.'

'And why so soon you go to Australia?'

'We're not from Australia. America.'

'Oh, America, I like. Rock and roll. Suzi Quatro. You live in California, yes?'

'How did you know?'

'All Americans from California. San Bernardino, okay? I meet doctor from there. With Polaroid. He took pictures. Was guest in my house.... Please wait! I show you! I am Datak.'

We stopped to see the smudged pictures and introduce ourselves, but we still weren't sure what he wanted.

'And where did you learn to speak English so well?'

'At university. You want to see university?' English was the universal language of wanting.

Iris looked at me, beginning to waver, but then Datak made a mistake. 'Very good batik work at university. Good prices. Friends to Datak.'

'We don't want good prices. We want to see Indonesia.'

'Okay, miss. No price for looking.... Indonesia very pretty.'

'Except for the army.' I couldn't resist a chance to gauge unrest among the student populace.

'Yes. Much army.' But Datak was merely concurring happily, enumerating another sight, and these three words were all the dissent I ever heard.

'Bye-bye!' Iris picked up the pace.

'All right. Goodbye. I think you come from book I read. American book.'

'What's that?'

'Yes. *Love Story*. All Americans know this book, yes? ... "Love, it is never having to not say you're sorry." Okay?'

Once we saw the buses crisscrossing in the station as if they were part of an unrehearsed Keystone chase, we were the ones who were sorry we didn't have Datak. But Iris was determined that we find our own way to the right mechanized sardine can....

'Prambanan!' she kept shouting to every wandering ticket-taker who could hear. 'Where in hell is Prambanana, you-all?' I half-expected to hear. If Iris couldn't find it, then it wasn't meant to be. But after accosting several touts, she got us on what seemed the right line. We leaned out the open front door all the way and were nearly thrown off with each swerve. At least, we got a good look at the glut of lopsided yellow-green hillocks, the midget bursts of banana grove and the royal palms above them all snooty. This was a volcanic golf course, a landscape in perspiration. At the right stop, the driver initiated a wave of elbowing that ended with our being shoved out the door. We were the only ones on the dirt road to the temple.

Past the usual row of Coke stands, we found four blackened turrets rotting on a high platform of grass above more rice paddies. At first, it looked like Prambanan was worthless without a good sand-blasting, but as we got closer, the grime took on a formidable order. Turrets and ramparts were lined with rungs of writhing, gesticulating figures: peewee warriors and midget goddesses who were bit actors in the long-running soap opera of the Ramayana saga. Gods turned to deer, deer to seducers, but these chiseled good guys and bad took their place in one fluid, three-hundred-and-sixty-degree strip.

To our amazement, the untended grounds were equipped with benches. Iris and I found one under a banyan tree and contemplated. Like the native foliage, Prambanan was excessively ornate, striving to attain the overdone. The medieval Javanese designers could never have belonged to the Bauhaus. Their handiwork put curlicues around curlicues even when the statement was as simple as devotion to duty. Where daily life was pared down, the imaginative life could not help but grow terrifyingly complex. The aesthetic of purposeful spareness could only flourish in glutted societies that were attempting to unencumber themselves. Less was *not* more, not here. More was more, less was little. Less than that was nothing at all.

An incomparable stillness invaded us on our bench. Was this the first knocking of the nonrational bogeyman? Or were our usual preoccupations just wilting in the heat? Prambanan struck us as holy because it was the first destination of ours that appeared untouched by the sordid exigencies of underdevelopment. At last, we were getting a glimpse of Indonesia. It was provided by this testament of a time when Indonesians weren't Muslims, weren't even Indonesians, when they didn't wish to be anything but what they were. Suddenly, we didn't wish them to be anything else, either.

Iris and I were captivated, but she didn't want to talk. Quickly, she grabbed my wrist and coaxed me into a game of hide-and-seek around the redolent stupas. We didn't exactly take turns giving chase—it was too hot for anything that enterprising and we didn't want to appear irreverent—so we just circled the sculpted bases, climbed the sacrificial steps from opposite sides of the towers, shouted 'Gotcha!' when we bumped heads in the shrines' mossy crawlspaces. We soon had a chance to play our game for real, when we spotted Datak, so cool in his Good Humor suit, breezing through the stubby portals of the sanctuary. Three Aussie lasses in sleeveless green dresses followed in tow, blond hair to their waists, camera straps shielding freckled, unviolated breasts. Obviously, he now had bigger foreign fish to fry, but the two of us did not want him to see that we'd lied to him and found this place without his aid. Escaping behind his back as he began an explication of the first Ramayana frieze, I wondered whether Datak could have told us anything more than what we'd intuited. What *World Book* facts could he have used to impress? Would he have brought forth more Erich Segal?

Buoyed by her first triumph, Iris insisted that we attend that night's performance of wayang kulit, the Javanese shadow play. 'And let's do it without some fool guide who'll explain it all to bits.' She shooed away offers from several more of those over dinner, asking only that they point the way. Each of them sent us toward the prince's candi. That seemed a logical enough setting for the venerated theatrical form, but we found no performance in progress there—only a lone guard who couldn't figure out what we were after. Beyond, in the entrails of old Yogya, all was discouragingly quiet. We wandered in a cobblestone maze, the high walls on either side dotted with alert-orange lichen. An occasional kerosene lamp twinkled, coaxing Iris on.

'Second time out, and you get us lost!'

'That's how I got kicked out of Girl Scouts.'

'You've got an infallible sense of direction. You're like a Geiger counter in reverse. . . .'

'Does it matter?'

'We should head back to town. We've probably missed half the show.'

'But it goes on all night!'

'Well, I don't.'

'Come on, crab. Look up there!'

Just before the next bend in the maze, a group of men had gathered and was staring up at something mounted high on one of their neighbor's stone balustrades. The attraction turned out to be the communal television, which was showing a subtitled rerun of 'I Dream of Jeannie'. Trying to follow the intrigue swirling about a Formica High School PTA meeting, the men were transfixed. None of them laughed at the situation comedy, and none of them knew anything about where we might find the shadow play.

'Just remember: the way back is always shorter than the way there.'

'Is that Iris' law?'

At our tenth dead end, three children pointed us three different ways. The last one led us to paydirt, where another internecine alley opened onto an oval dirt lot, the kind where operators back home might have staged a demolition derby. Thankfully, there was nothing here to demolish and the tracks in the lot were made by the thinner trade of a hundred or so vendors with their wooden carts. Over each vat of frying shrimp, rice or dough, wobbled a kerosene lamp or several candles. The aggregate effect was a production number from *Peter Pan*: more fireflies, pure candi-land. At the far end of this tremulous night market and nymphs' convention was the one structure in Jogjakarta that made a stab at indestructibility. Its style was congressional Greco-Roman. Rounded steps took us inside an auditorium that might have belonged to any decent-sized American high school.

It wasn't so easy to tell if the shadow play had already begun. The house lights were on, children in baggy shorts romped everywhere. But a triumphant Iris led us toward two of the many empty folding seats. We had conquered the tropics, we had stumbled on culture—and, up ahead, at floor level, was the portable proscenium of the wayang kulit.

On the front side of the two side flaps of a rice paper screen hung the several dozen puppets used to cast the shadows. No attempt was made to hide this company in the wings or create some Punch 'n Judy illusion. These solid forms, like ours, were deceptive. Reality was in the reflections, natch. These gods on folding skewers were representatives of the universe's controlling forces, and it didn't matter how fragile they looked. Local dramaturges had discovered long ago that what made the gods believable was the audience's belief. This theater was all the more epic because the viewers did most of the work. The concert-master remained unhidden and attempted no disguise of his manipulation. The puppets were hauled off the rack one at a time and held before a light bulb at center stage. They were thrust so close to the screen that hardly any shadows were cast at all, though the figures grew more spectral as the light shone through their gossamer hides. With their sneering, elongated chops, Jimmy Durante schnozzes, each of the characters in this eloquently deformed cast appeared to be flinging down curses upon their enemies, and us. Their headgear nearly crisped up with the flame of their righteousness. The puppets' attenuated grasshopper limbs brought to mind an eternal starvation—that permanent advantage known as lack of want—but the arms were their only moving parts. Needed to point, accuse and condemn. The man who controlled these few stilted gestures was also the principal singer. Kneeling beside the characters he animated, this impresario took all parts, muttering with an operatic cadence into a knee-high microphone. Again, artifice was shunned. Special effects were up to the viewer. The gods raged and were frighteningly well-defined, perhaps the most specifically imagined demons on earth, but their setting was minimal, their world static. Indeed, their mythical deeds took place right in our laps.

Was that why each performance took up to sixteen hours? It was hard to imagine that a single concert-master could last through the whole script, but this one had no trouble with speeches that droned on for half an hour, and we couldn't spot any understudies in the wings. But he got a break, as did the audience, at the end of each fierce tableau. An interlude was provided by a gamelan orchestra that separated stage from audience on the teak floor. No pit hid these musicians in their resplendent checkered sarongs and matching forehead wraps. Their instruments, too, were part of the spectacle, from gongs that could have been prized ceramic vases to sound-boxes that resembled sets of pachyderm ribs. Playing as one, the herky-jerk of this xylophonic ensemble resonated out of time and into our ears. Atonal modernists had prepared us to appreciate this most ancient attempt to give chaos a beat and magic precision. The men who sat on the floor before us, tapping their strikers and asserting the primacy of the bell, were making their music as stand-ins for the wind.

The Javanese did not tap their toes, or react overtly to lyrics or score. Yet this was clearly a big event for the town. The seats at the front were hogged by dignitaries, all male, who wore black felt caps shaped like the paper ones worn by managers at a McDonald's franchise. These bespoke a more enduring authority, as well as their Moslem faith. Throughout the performance, Jogjakarta's Chamber of Commerce kept their arms folded, most seriously, over bellies that were their most conspicuous show of wealth. They also did their best to ignore the noisy and irreverent peanut gallery behind them. Not only did the commoners chat away, but their pregnant women fidgeted and ambled about, their barefoot children played with string or leapfrogged the empty chairs. In this auditorium, there were few of the trappings that accompany Western 'high art': no overzealous aficionados or determination to appreciate, no snobbery or diamond brooches or handy programs at the door. The shadow play was actually play, not some hallowed cultural expression to be endured. It was also historic, but since history's continuum was relatively unbroken in these parts, there was no reason the play had to be revered. It was an old friend that everyone took for granted.

There could have been little forced solemnity, since the audience was mostly children. It was a sight to which we'd soon become accustomed: always enough children to outnumber their elders, always more hands clutching at fewer resources. Just don't tell that to the kids of Jogjakarta. They were getting a full measure of entertainment from their 'night out'. Many were unaccompanied, and the rest were strikingly unleashed, joyfully tolerated. Darting and squirming around and over the folding seats, their black hair uniformly shaved, this local gang looked like monks who'd taken a vow of cuteness. The shape of their skulls seemed equally anti-Western. They came to no points, were not sharpened at the cerebrum, but remained unbiased toward the rational, rounded on top and bulging at the sides like squashed fruits. They all seemed stuck at that age where their heads were too big for their bodies. In turn, the heads could not contain their eyes: such perfect spoons of white pigment.

Many of those eyes were drawn to a rotund German who'd plunked down in their midst and was beckoning all the boys. When they weren't climbing all over his considerable terrain, or returning playful slaps, or bouncing on his ham hock knees, we could see he was dressed only in an orange sarong girdle around his middle. This son of Saxony had gone thoroughly Gauguin, though that only revealed how luridly pale and blue-veined he was beside all that *kinderfleisch*. The rest of the audience paid him no heed, didn't seem to notice the vague prurience in his beaming. Never had this man had an opportunity to so fully ooze with paternalism, to fondle so many pups. None of his changelings seemed to mind that he stroked their shaved domes as if trying to polish brass knobs: they didn't know that a pat on the head carried a price.

It was hard to keep our eyes on the concert-master's dexterity instead of the German's. But we had a train to catch in the morning and walked out on the third act, which was indistinguishable to our eyes from second and first. Was that why the Javanese went on gabbing throughout? Sneaking away and down the front steps of the auditorium, I was about to hail one of the hundred waiting rickshaws—no doubt, they, too, would be there until the play ended at dawn. But Iris flinched and tugged at my arm. Standing in the lot, front row center, before the aisles of shaky pushcarts and quivering candelabras, was our nemesis, the shadow we cast.

'Australian?' The approach began once again. But, no—this wasn't Datak, just another of his lanky cousins, who looked just as sporting, and as clownish, in pleated white pants.

'American.'

'America! Long way to Jogjakarta!'

'Yes, long way.'

'Yes see shadow play?'

We nodded. What else?

'You see Prambanan temple?'

We nodded again.

'You see my house?'

'Thank you, no.' Was it one of the local treasures? 'Catch train tomorrow.'

'I know. You go Bali.... But first, you be friends?'

'Why?' Iris asked. It sounded like an awfully dumb question, but she was learning through imitation.

'Friendship good. Understanding. Then we take pictures. Then you buy batik.'

Perhaps there was some friendship in such an interchange, but for us, it too resembled a kind of mutual bondage, the boys on the great white father's lap.

'No, we go to sleep now.'

'All right. Very good. I see you wish to be alone. I see you are very much in love.'

'We are?' Iris asked again.

'Yes, yes. In Indonesia, we call love "the meeting feeling".' And our shadow said goodbye to us by illustrating the phrase with a clasping of nimble-jointed, tweezer-shaped, marionette fingers.

'You see,' I told Iris, my arm around her in the becak, the recitatives of the real shadow play fading, 'we're fooling the whole world!'

Later that night, we were most strenuously reminded of how much we were a part of the 'whole world'. An earthquake, brief but insistent as a cough, made Iris and me sit up in the dark. Blinded, we could hear new fissures being released in the loesman plaster, the teeth of tin roofs chattering, and the women who slept on blankets along the boardwalk raise up a wail. They did not seem to be baying so much from fear as out of nocturnal arousal. These ladies of Java couldn't be afraid, I reasoned, half-swallowed by the ground as they already were. They didn't have such a long way to fall.

John Krich, *Music in Every Room: Around the World in a Bad Mood*, New York: Atlantic Monthly Press, 1984, pp. 61–72. Reprinted with the permission of McGraw-Hill, Inc.

35
Richard Critchfield Goes Back to Java, 1985

'Hello, mister! Where are you going?'[1] *Thus did pedicab driver Husen announce himself to journalist Richard Critchfield in 1967. Husen went on to introduce Critchfield to the life of Jakarta's working poor and to the village world as well. Critchfield came to know Husen and his family circle and, in subsequent visits over the years, chronicled Husen's rise from poverty and the permutations in his domestic life, all against the backdrop of profound social and economic changes on Java. Here he reflects on some of these changes as of 1985.*

ONE tends to think of countries, of cultures, in terms of the individuals one knows best. In Java, to me this means Husen, a *betjak* or three-wheeled-pedicab driver, who introduced me first in 1967, and many times since, to village life.

A Javanese village can be steamy and the equatorial heat unbearable, especially out in a rice paddy at midday. But in the early morning or late afternoon, when a breeze blows in from the Java Sea, it can also be very pleasant. Against a backdrop of misty volcanoes, blue and hazy, placid muddy river, bright green paddies and clumps of bamboo, mangoes, and bananas, the steamy days drift into balmy nights. After dark, as oil lamps are lit (only 15 percent of Java's villages have electricity though many peasants have Japanese 'gen-sets' to provide their own), one usually hears the haunting *ning-nong* sound of bamboo gamelan orchestras floating over from some distant treeline. At night the countryside is alive with shadow plays, folk dramas, classical dance and acrobatic performances, all put on by the villagers themselves. Together with the Balinese, the Javanese are possibly the most cultured, artistic, and mystical villagers anywhere....

Husen's village of Pilangsari, over the years, has prospered. In 1967, when I first visited it, most of the villagers lived in

[1]Richard Critchfield, *The Golden Bowl Be Broken*, Bloomington, Indiana: Indiana University Press, 1973, p. 223.

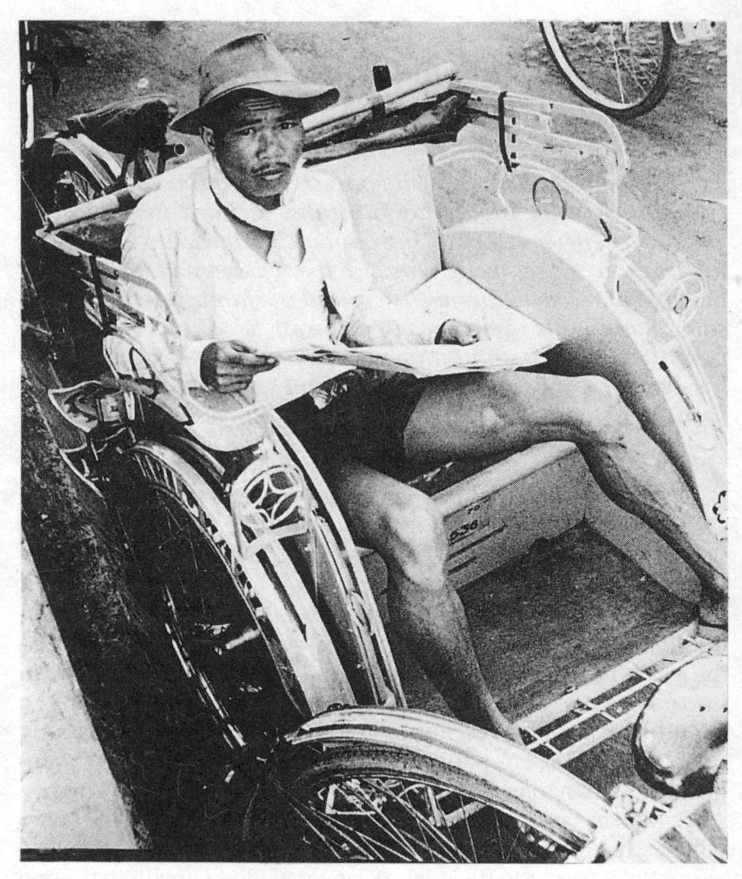

'Hello, mister!' Critchfield's friend Husen rests in his *betjak*. (From Richard Critchfield, *The Golden Bowl Be Broken*, Bloomington, Indiana: Indiana University Press, 1973)

bamboo huts with earthen floors, a coconut-fiber roof and often no windows. Everyone slept on mats on raised wooden platforms, used oil lamps, and took water from a spring or river; there were few bicycles or radios. Today, thanks to the rice revolution, most of the villagers have brick houses with tile roofs, polished tile floors, and glass-paned louver windows; they light their rooms with electricity or a petrol lamp. They sleep in beds with mosquito nets, draw water from individual wells, have outdoor privies, and may own a motor scooter, television set, or even a car.

In the late 1960s and early '70s, the only practicing Muslims, or *santri*, whom I knew in the village were Husen's father and Abu, a student from Bandung who worked as an agricultural extension agent, teaching the villagers about high-yield rice and how to grow it. Husen's parents strongly resisted change. I think they suspected, probably rightly, that if they gave in at all to innovations, there was no telling where it would stop. Finally, in 1973, his father, at Husen's urging, planted his first crop of Philippine-bred high-yield dwarf rice, investing in nitrogen fertilizer and chemical insecticide for the first time. He tripled his normal yield, harvesting six tons on his two-and-a-half acres of rice land. But as soon as we left, Husen to pedal his betjak in Jakarta and I for another country, the father returned to traditional methods, scattering only leaves from the *jowar* tree on the green shoots of his next crop and relying on prayers and fasting to keep insects away. The harvest fell back to two tons.

It was not until a visit in 1978 that I found everybody, including Husen's father, growing fertilizer-intensive high-yield rice. (Indeed, the father seemed embarrassed when I recalled how bitterly he had opposed it.) Average per-hectare yields were six tons; a few enterprising villagers were getting as much as ten. An old Dutch-built canal, repaired in 1974, provided year-round water; there were double or triple rice crops for everybody.

Pilangsari had a new school. (One of Suharto's real achievements has been to build a primary school in almost every Indonesian village. In 1985 primary education was made universally compulsory, though there is an extremely high drop-out rate after three grades and adult literacy in Indonesia is just 68 percent.) Hondas were replacing bicycles. New Japanese buses and trucks rushed people at frightening speeds along newly paved highways to market towns. Husen had left work in the city—he had graduated from pedaling a betjak to driving a taxi—and was now farming full time. His father had built a seven-room brick house. 'It's his monument,' Husen said.

The old traditions seemed to be holding their own. Old women who planted and harvested rice were still getting one-sixth of the crop. There seemed to be as many wayang performances as ever. 'We have come onwards,' Husen told me in English with a grin.

Five years went by. In 1981 Husen wrote to tell me of the death of his father. Some months later he moved to the village of a new, fourth wife, a two-hour journey by bicycle or a half-hour by bus from Pilangsari. The new village, Kedungsari, was on higher ground, unirrigated, and much poorer.

Even before the father's death, the Citanduy River had eaten away some of his land. A World Bank-funded project to control the river's flow chopped off some more. When he died, his land had shrunken from 700 batas, equivalent to one hectare, down to about 500 batas. Counting his mother and eight siblings, Husen's share was just 50 batas, too small a plot of land to make his livelihood. In Kedungsari, where land was cheaper, he was able to buy a quarter-hectare and was saving to buy another quarter.

Kedungsari was off the beaten path, some seven kilometers from the nearest main road. As elsewhere in Java, the deeper you went into the hinterland, the more the bicycles, motor scooters, and cars, and modern, tile-roofed bungalows, began to disappear. The land was still lush and green, but there was a sense of moving back in time. In the village, while there was a new primary school, almost all the homes were windowless, thatched bamboo huts set in their unirrigated gardens of bamboo, banana, and mango trees.

In 1985, Husen, now 47, seemed more content than at any time in the 18 years I had known him. He seemed to have made his peace. 'I am very happy as a farmer. There is good air here. It is quiet. If out in the village one can look far across the country. In Jakarta there were always buildings.'

It was a theme he returned to again and again. 'For me, I was looking for a quiet place, not busy. And quiet and good air. About other people, I don't know. If for me, already stay in Jakarta, I don't like. Better in the village. But must work really hard. If very lazy, the village is no good.'

He said very few people from Kedungsari migrated to Jakarta, just young men who went for three or four months to see what it was like and then came home again. He found the old village tradition of *gotong royong*, mutual help, much stronger than in Pilangsari where, with growing prosperity, there had been a shift to material values.

Life had become 'so empty' in Jakarta, 'not quiet like now'. He said, 'In the village we are like ants getting the food, little by little, if long time they make a hill, working hard.' People were more neighborly in Kedungsari. 'If I don't have rice, the neighbors give and help. Gotong royong is very nice. Before the road was muddy, motor car cannot [go] inside. Everybody carried bricks and stones from the river and made the road good. Now in Pilangsari gotong royong not so good like before.'

Was development bad then? 'No, no,' Husen protested. 'I am also like to go up, not like to come down, everybody likes to prosper.'

The biggest surprise was that Husen was now santri [a devout Muslim]. He practiced a purer Islam, carefully observing the five prayers a day, he went to the village mosque, in Ramadan he had fasted from dawn to sunset. In all the years I'd known him, he had done none of these things before. It wasn't only him, he said, his brothers, Waryono and Marjo, his sisters, Sutini and Ermina, and so many of their friends and neighbors in both Pilangsari and Kedungsari had become santri.

Why? Husen could give no reason. When I persisted asking, he finally said, 'If motor car, you have brake.' Ah, when change came too fast, you turned to Islam?

What about Semar [Hindu god-clown of the *wayang*]? There had been, it turned out, no evident turning away from the wayang. Indeed, Husen had helped his fellow villagers in Kedungsari collect 300,000 rupiah at harvest time (not quite $300) to hire a dalang [puppeteer] to perform *Mapag Sri*, a story specially performed in the Cirebon region of Java, where Husen lives, to celebrate the rice harvest.

During the time I spent in Kedungsari, Husen and I went about the countryside for hours at a time by bicycle. Monsoon skies, when it is not actually raining, can be intensely blue, with row after row of white, thick cumulus clouds, drifting over the wet, pale-green surfaces of the rice paddies. While we mainly stuck to narrow dirt tracks, even on a paved road there was little traffic but the occasional woman on a bicycle, carrying fruit to market, or a few horse-drawn carriages. In many fields people were planting spice (as Husen himself had done just before I came). The only sign of change was a man plowing a field with a power tiller (he said it cost $3,000 and one needed at least a hectare of land to own one). Gone was the heavy traffic of the main roads, with their speeding buses and trucks, roaring scooters, and smoking haze of black exhaust.

Kedungsari, poor as it was, with its thatched huts of woven bamboo and small concrete terraces, was like Java the first time I saw it. Some of the houses still had shrines for Dewi Sri [rice goddess] behind the pillared main room. Husen felt comfortable with the security of the old traditional village of shared poverty, communal labor, the obligatory night watch. It was remote and small and poor enough to escape most 'development', and there was still about it a sense of the old enchantment, as if the range of volcanic mountains that rose in the distance had held the world at bay.

Why, after bustling, prospering Pilangsari, had Husen chosen to come here? (He said it was his wife's wish; it was either come or divorce. This I found hard to believe and suspected it was not the true reason.) Why, also, had he become a 'good' Muslim? Husen, of all people, who had so exemplified the peasant culture, nominally Muslim but Hindu in many of its deepest values and also animist and mystical. When he said 'If motor car, you have brake', what did he mean?

Richard Critchfield, 'Going Back to Java', *Universities Field Staff International (UFSI) Reports*, 1985, No. 32, pp. 4–7.

Other Oxford Paperbacks for readers interested in South-East Asia, past and present

Cambodia

Angkor: An Introduction
GEORGE COEDÈS

Angkor and the Khmers
MALCOLM MacDONALD

Indonesia

An Artist in Java
JAN POORTENAAR

Bali and Angkor
GEOFFREY GORER

Coolie
MADELON H. LULOFS

Diverse Lives: Contemporary Stories from Indonesia
JEANETTE LINGARD

Flowering Lotus: A View of Java in the 1950s
HAROLD FORSTER

Forgotten Kingdoms in Sumatra
F. M. SCHNITGER

The Head-Hunters of Borneo
CARL BOCK

The Hidden Force*
LOUIS COUPERUS

The Hunt for the Heart: Selected Tales from the Dutch East Indies
VINCENT MAHIEU

In Borneo Jungles
WILLIAM O. KHRON

Island of Bali*
MIGUEL COVARRUBIAS

Java: Facts and Fancies
AUGUSTA DE WIT

Java: The Garden of the East
E. R. SCIDMORE

Java: A Travellers' Anthology
JAMES R. RUSH

Javanese Panorama
H. W. PONDER

The Last Paradise
HICKMAN POWELL

Let It Be
PAULA GOMES

Makassar Sailing
G. E. P. COLLINS

The Malay Archipelago
ALFRED RUSSEL WALLACE

The Outlaw and Other Stories
MOCHTAR LUBIS

The Poison Tree*
E. M. BEEKMAN (Ed.)

Rambles in Java and the Straits in 1852
'BENGAL CIVILIAN' (C. W. KINLOCH)

Rubber
MADELON H. LULOFS

A Tale from Bali*
VICKI BAUM

The Temples of Java
JACQUES DUMARÇAY

Through Central Borneo
CARL LUMHOLTZ

To the Spice Islands and Beyond: Travels in Eastern Indonesia
GEORGE MILLER

Travelling to Bali
ADRIAN VICKERS

Twin Flower: A Story of Bali
G. E. P. COLLINS

Unbeaten Tracks in Islands of the Far East
ANNA FORBES

Witnesses to Sumatra
ANTHONY REID

Yogyakarta: Cultural Heart of Indonesia
MICHAEL SMITHIES

Malaysia

Among Primitive Peoples in Borneo
IVOR H. N. EVANS

An Analysis of Malay Magic
K. M. ENDICOTT

At the Court of Pelesu and Other Malayan Stories
HUGH CLIFFORD

The Best of Borneo Travel
VICTOR T. KING

Borneo Jungle
TOM HARRISSON

The Chersonese with the Gliding Off
EMILY INNES

The Experiences of a Hunter and Naturalist in the Malay Peninsula and Borneo
WILLIAM T. HORNADAY

The Field-Book of a Jungle-Wallah
CHARLES HOSE

Fifty Years of Romance and Research in Borneo
CHARLES HOSE

The Gardens of the Sun
F. W. BURBIDGE

Glimpses into Life in Malayan Malayan Lands
JOHN TURNBULL THOMSON

The Golden Chersonese
ISABELLA BIRD

Illustrated Guide to the Federated Malay States (1923)
C. W. HARISSON

The Malay Magician
RICHARD WINSTEDT

Malay Poisons and Charm Cures
JOHN D. GIMLETTE

My Life in Sarawak
MARGARET BROOKE, THE RANEE OF SARAWAK

Natural Man
CHARLES HOSE

Nine Dayak Nights
W. R. GEDDES

A Nocturne and Other Malayan Stories and Sketches
FRANK SWETTENHAM

Orang-Utan
BARBARA HARRISSON

The Pirate Wind
OWEN RUTTER

Queen of the Head-Hunters
SYLVIA, LADY BROOKE, THE RANEE OF SARAWAK

Six Years in the Malay Jungle
CARVETH WELLS

The Soul of Malaya
HENRI FAUCONNIER

They Came to Malaya
J. M. GULLICK

Wanderings in the Great Forests of Borneo
ODOARDO BECCARI

The White Rajahs of Sarawak
ROBERT PAYNE

Philippines

Little Brown Brother
LEON WOLFF

Singapore

The Manners and Customs of the Chinese
J. D. VAUGHAN

Raffles of the Eastern Isles
C. E. WURTZBURG

Singapore 1941–1942
MASANOBU TSUJI

Travellers' Singapore
JOHN BASTIN

South-East Asia

Adventures and Encounters
J. M. GULLICK

Adventurous Women in South-East Asia
J. M. GULLICK (Ed.)

Explorers of South-East Asia
VICTOR T. KING (Ed.)

Soul of the Tiger*
JEFFREY A. McNEELY and PAUL SPENCER WACHTEL

Thailand

Behind the Painting and Other Stories
SIBURAPHA

Descriptions of Old Siam
MICHAEL SMITHIES

The English Governess at the Siamese Court
ANNA LEONOWENS

The Politician and Other Stories
KHAMSING SRINAWK

The Prostitute
K. SURANGKHANANG

Temples and Elephants
CARL BOCK

To Siam and Malaya in the Duke of Sutherland's Yacht *Sans Peur*
FLORENCE CADDY

Travels in Siam, Cambodia and Laos 1858–1860
HENRI MOUHOT

Vietnam

The General Retires and Other Stories
NGUYEN HUY THIEP

The Light of the Capital: Three Modern Vietnamese Classics
GREG & MONIQUE LOCKHART

Titles marked with an asterisk have restricted rights.